The Dynamics of Meristem Cell Populations

ADVANCES IN EXPERIMENTAL MEDICINE AND BIOLOGY

Volume 1
THE RETICULOENDOTHELIAL SYSTEM AND ATHEROSCLEROSIS
Edited by N. R. Di Luzio and R. Paoletti • 1967

Volume 2
PHARMACOLOGY OF HORMONAL POLYPEPTIDES AND PROTEINS
Edited by N. Back, L. Martini, and R. Paoletti • 1968

Volume 3
GERM-FREE BIOLOGY: Experimental and Clinical Aspects
Edited by E. A. Mirand and N. Back • 1969

Volume 4
DRUGS AFFECTING LIPID METABOLISM
Edited by W. L. Holmes, L. A. Carlson, and R. Paoletti • 1969

Volume 5
LYMPHATIC TISSUE AND GERMINAL CENTERS IN IMMUNE RESPONSE
Edited by L. Fiore-Donati and M. G. Hanna, Jr. • 1969

Volume 6
RED CELL METABOLISM AND FUNCTION
Edited by George J. Brewer • 1970

Volume 7
SURFACE CHEMISTRY OF BIOLOGICAL SYSTEMS
Edited by Martin Blank • 1970

Volume 8
BRADYKININ AND RELATED KININS: Cardiovascular, Biochemical, and Neural Actions
Edited by F. Sicuteri, M. Rocha e Silva, and N. Back • 1970

Volume 9
SHOCK: Biochemical, Pharmacological, and Clinical Aspects
Edited by A. Bertelli and N. Back • 1970

Volume 10
THE HUMAN TESTIS
Edited by E. Rosemberg and C. A. Paulsen • 1970

Volume 11
MUSCLE METABOLISM DURING EXERCISE
Edited by B. Pernow and B. Saltin • 1971

Volume 12
MORPHOLOGICAL AND FUNCTIONAL ASPECTS OF IMMUNITY
Edited by K. Lindahl-Kiessling, G. Alm, and M. G. Hanna, Jr. • 1971

Volume 13
CHEMISTRY AND BRAIN DEVELOPMENT
Edited by R. Paoletti and A. N. Davison • 1971

Volume 14
MEMBRANE-BOUND ENZYMES
Edited by G. Porcellati and F. di Jeso • 1971

Volume 15
THE RETICULOENDOTHELIAL SYSTEM AND IMMUNE PHENOMENA
Edited by N. R. Di Luzio and K. Flemming • 1971

Volume 16A
THE ARTERY AND THE PROCESS OF ARTERIOSCLEROSIS: Pathogenesis
Edited by Stewart Wolf • 1971

Volume 16B
THE ARTERY AND THE PROCESS OF ARTERIOSCLEROSIS: Measurement and Modification
Edited by Stewart Wolf • 1972

Volume 17
CONTROL OF RENIN SECRETION
Edited by Tatiana A. Assaykeen • 1972

Volume 18
THE DYNAMICS OF MERISTEM CELL POPULATIONS
Edited by Morton W. Miller and Charles C. Kuehnert • 1972

The Dynamics of Meristem Cell Populations

The Proceedings of a conference jointly organized by the Department of Radiation Biology and Biophysics, The University of Rochester, and the Department of Biology, Syracuse University, and convened at Rochester, New York, August 19-21, 1971.

Edited by
Morton W. Miller
Department of Radiation Biology and Biophysics
School of Medicine and Dentistry
The University of Rochester
Rochester, New York

and

Charles C. Kuehnert
Department of Biology
Syracuse University
Syracuse, New York

PLENUM PRESS • NEW YORK–LONDON • 1972

Library of Congress Catalog Card Number 70-185045

ISBN 978-1-4684-3209-1 ISBN 978-1-4684-3207-7 (eBook)
DOI 10.1007/978-1-4684-3207-7

A Division of Plenum Publishing Corporation
227 West 17th Street, New York, N.Y. 10011

United Kingdom edition published by Plenum Press, London
A Division of Plenum Publishing Company, Ltd.
Davis House (4th Floor), 8 Scrubs Lane, Harlesden, NW10 6SE, London, England

PREFACE

These proceedings are composed of three parts: 1) the individual papers, 2) their discussions, and 3) a retrospective summation of the meeting's deliberations. If we were to suggest a starting place for reading it would probably be the summation -- it attempts to place the various papers and discussions into a general context and to point out relevant ideas of current knowledge and of potentially useful avenues for further research. The papers and discussions deal more specifically with the individual topics.

It is hoped that the tone and tenor of the meeting are somewhat reflected in its proceedings. The emphasis today in biological research is quite "ecumenical" in outlook in that often techniques and conclusions from one discipline have relevance and applicability to another. For example, the elucidation of the DNA synthetic stage during interphase was first accomplished with botanical material; this technique has since had a broad and significant role in medical research. Also, the pioneering and now classical studies regarding the effects of ionizing radiation on plant chromosomes were, and still are, the foundation for studying mammalian chromosomes: it required only the discovery of a hypotonic medium to achieve spreading of them so that similar types of analyses were possible. Plants and plant cells have many unique features which allow them to be useful either for the study of specific botanical problems, or for studying broad biological principles.

When dealing with a single broad topic it is often very useful for scientists to meet and discuss their work in a relatively closed seminar type of setting. This was the format chosen for this meeting. The goal of the meeting was to achieve a better understanding of meristem organization and function in relation to meristems *per se* or in relation to uses of meristems for elucidating fundamental biological principles (i.e., in those instances where the meristem is the "tool" and not necessarily the subject). Though the scientists present at these proceedings variously considered themselves as morphologists or as anatomists, as cell biologists or as radiation biologists, nevertheless, each of them utilized a plant meristem as a tool. What resulted was something akin to the viewing of a multifaceted crystal, with different "angles" giving different colors, different types of illumination. It was inevitable that conflicting opinions should exist; it is hoped that they are reflected in the contents of this volume. In a meeting where the expertise of the audience is no less than that of the speakers, the discussions are frequently as valuable as the papers themselves.

The papers in this volume constitute a broad, but well-balanced portrayal of meristems both as subjects and as tools. The papers contain comprehensive coverage of many new and important developments with a bounty of new data and formulations. The discussions following each paper, and the conference summation, are in some fashion further amplification, further integration, and further interrogation of the information "bits" contained in the papers. Clearly, plant meristems have been and will continue to be useful vehicles for a variety of reserach endeavors. This assessment of meristems from a variety of viewpoints hopefully achieves a greater understanding of meristem function.

The Editors

CONTRIBUTORS

Ernest A. Ball
Department of Developmental and Cell Biology
University of California
Irvine, California

F. A. L. Clowes
Botany School
Oxford University
Oxford, England

Alan D. Conger
Department of Radiobiology
Temple University
Philadelphia, Pennsylvania

Bob V. Conger
AEC Agricultural Research Laboratory
The University of Tennessee
Oak Ridge, Tennessee

Milton J. Constantin
AEC Agricultural Research Laboratory
The University of Tennessee
Oak Ridge, Tennessee

Elizabeth G. Cutter
Department of Botany
University of California
Davis, California

Francesco D'Amato
Institute of Genetics
The University
Pisa, Italy

Douglas Davidson
Department of Biology
McMaster University
Hamilton, Ontario, Canada

E. A. Favret
Department of Agronomy and Soils
Washington State University
Pullman, Washington

Donald E. Fosket
Department of Developmental and Cell Biology
University of California
Irvine, California

Alan H. Haber
Biology Division
Oak Ridge National Laboratory
Oak Ridge, Tennessee

Barrie E. Juniper
Botany School
Oxford University
Oxford, England

D. D. Killion
AEC Agricultural Research Laboratory
The University of Tennessee
Oak Ridge, Tennessee

Calvin F. Konzak
Department of Agronomy and Soils
Washington State University
Pullman, Washington

C. J. Kovacs
Biology Department
Brookhaven National Laboratory
Upton, New York

Charles Carroll Kuehnert
Department of Biology
Syracuse University
Syracuse, New York

Morton W. Miller
Department of Radiation Biology and Biophysics
The University of Rochester
Rochester, New York

Thomas S. Osborne
Department of the Biological Sciences
Smith College
Northampton, Massachusetts

John R. K. Savage
Medical Research Council
Radiobiology Unit
Harwell, England

Harold H. Smith
Biology Department
Brookhaven National Laboratory
Upton, New York

John G. Torrey
Harvard Forest
Harvard University
Petersham, Massachusetts

Jack Van't Hof
Biology Department
Brookhaven National Laboratory
Upton, New York

CONFERENCE PARTICIPANTS*

Ying Kai Fu
Department of Radiation Biology and Biophysics
The University of Rochester
Rochester, New York

Winborn Gregory
Department of Radiation Biology and Biophysics
The University of Rochester
Rochester, New York

Martha Hedden
Winthrop College
Rock Hill, South Carolina

Edward H. Jordan
Department of Biology
Syracuse University
Syracuse, New York

Gary E. Kaufman
Department of Radiation Biology and Biophysics
The University of Rochester
Rochester, New York

Christopher S. Lange
Department of Radiation Biology and Biophysics
The University of Rochester
Rochester, New York

Haviva Langenauer
Department of Biology
Smith College
Northampton, Massachusetts

Christopher W. Lawrence
Department of Radiation Biology and Biophysics
The University of Rochester
Rochester, New York

Daniel Lieberman
Department of Radiation Biology and Biophysics
The University of Rochester
Rochester, New York

*In addition to contributors

CONFERENCE SECRETARY

Nancy A. O'Brien
Department of Radiation Biology and Biophysics
The University of Rochester
Rochester, New York

ACKNOWLEDGEMENTS

Partial support for this conference was gratefully provided by the United States Atomic Energy Commission, Washington, D.C. and Educational Modules, Inc., Rochester, New York. The conference committee wishes to express their appreciation to Dr. John R. K. Savage for his editorial consultancy in preparing the Discussions of each paper, to Drs. Douglas Davidson and Barrie Juniper for their time and effort in editing the proceeding's Discussions, to Nancy O'Brien for her patient and efficient secretarial service, to Ying Kai Fu for his most efficient operation of the projection equipment, and to Martha Hedden for lending a helping hand throughout. The conference committee is especially indebted to every participant for the uniform high quality and attention to format of his paper, and to the session chairmen for their able and enthusiastic handling of their assignment. We also wish to acknowledge with gratitude the services of Cyril Lichtensteiger of the International Recording Guild, Staten Island, New York, for his superb recording of the Discussions which so eased the task of editing. And finally, we thank our wives Marylynn and Bertha, who have been most patient and helpful.

Morton W. Miller
Charles Carroll Kuehnert

CONTENTS

SESSION V

GENERAL DISCUSSION AND SYNTHESIS; CONCLUSIONS

Alan D. Conger, Chairman

The Dynamics of Meristem Cell Populations

ON THE INITIATION OF ORGANIZATION IN THE ROOT APEX*

JOHN G. TORREY

Harvard Forest, Harvard University

Petersham, Massachusetts, 01366

INTRODUCTION

Root and shoot apices are the major centers of organization in the development of the higher vascular plant. Apices are sites of origin of new cells formed by repeated cellular divisions in distinctive patterns; they serve as the organizers for the elaboration of tissue patterns in the continuing development of the mature axial structures of the root and stem. Apices are where things happen. Once formed, an apex, either root or shoot, is a remarkably stable structure, unperturbable in its activities of new cell formation and tissue pattern determination. Furthermore, it hides the secrets of its inherent organizational capacity remarkably well. We know very little about the properties of apices which allow them to be essentially autonomous in their functioning. Their activities are programmed in the genetic constitution of the cells of which they are comprised and the coordinated, multicellular tissue patterns which result from correlated intercellular behavior are equally referable to the genes. One of the basic problems of understanding organization in the vascular plants is to understand how apices operate, in what ways they are themselves organized and how this distinctive organization arises.

The root apex is the simpler system to study. Its basic organization is devoted to the formation of the axial structure of the root itself, initially uncomplicated by lateral appendages. Only the bidirectional production of cells--for construction of the root proper in the proximal direction and for the root cap in the distal direction--complicates the structure of the apex.

Experimental analysis of root structure is facilitated by its terminal position on an underground structure. The root is heterotrophic, dependent upon carbohydrate, certain vitamins and perhaps hormones translocated to the tip from the photosynthetic shoot. It can be grown in organ culture bathed in a relatively simple nutrient medium in the dark where it continues its precisely patterned multicellular activities.

EVIDENCES OF ORGANIZATION

Analysis of the organization of the root apex has been accelerated remarkably in the last two decades by the study of subpopulations of cells in the apex. In 1956 Clowes[6] first reported the presence of a population of cells in the root apex of Zea which was distinctive in having cells which incorporated labelled precursors of DNA very little if at all if the precursors were provided for the duration of an average cell cycle. The population, which Clowes called the quiescent center, occupied a subterminal position in the root tip, had roughly the shape of a hemisphere with the flat surface of the half-sphere bordered by the root cap and specifically centered over the cells of the columella of the root cap.

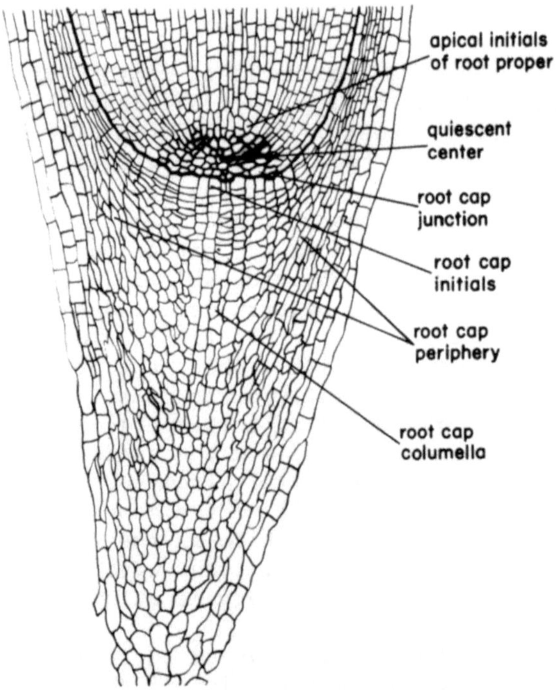

Figure 1. Diagram of median longitudinal section of root tip of Convolvulus showing the regions of the apex (from Phillips and Torrey[20]).

In intensive studies of this regional center Clowes and his associates showed that the cells of the quiescent center have very prolonged cell cycle times compared to the rest of the cells of the apex. In fact, several subpopulations of cells with various cycle times can be defined around the quiescent center. In Figure 1 are illustrated the different cell populations seen in the Convolvulus root as determined by Phillips and Torrey[18, 19, 20], using the method of metaphase accumulation following colchicine treatment. The values for cycle times in Convolvulus are similar to published reports by Clowes and others, using this method and the method of pulse labelling with ^{3}H-thymidine, and confirm this type of sub-population distribution reported for root apices of a number of different plants. In Table I are summarized the published data for these sub-populations. According to evidence obtained from DNA measurements of cells in the root apex (Clowes[11, 12]), cell populations differ from each other primarily in having differing lengths of the G_1 phase of the cycle, i.e., the period preceding the synthetic period for DNA. Thus the cell populations are fundamentally different from each other primarily with respect to one part of the cell cycle, that is, the rate at which they progress through G_1. In the quiescent center itself, a mechanism exists within the cells which arrests the cell cycle for a prolonged period, preventing the usual entry into S and DNA synthesis. And it is through the control of the duration of G_1, or perhaps more specifically the entry into the S period of the cells located in various regions of the root apex that the organization of the apex is determined. The most actively dividing cells are the apical initials. There exist several apical initial groups which may be identified anatomically in relation to their cell progeny and physiologically in relation to their cycle times. Two sets of initials give rise to the root cap, viz., the root cap initials distal to the quiescent center which form primarily the cells of the root cap columella. These initials have a very short average cell cycle time and form a constantly renewed cell population (Phillips and Torrey[19]). A less clearly defined group of initials forms the peripheral root cap. On the proximal surface of the quiescent center arranged in the shape of a hemisphere are the initials of the root cortex and central cylinder which are most easily subdivided according to the tissue systems they form. They also have relatively short cycle times and account for the bulk of the axial structure of the root. Central to all of these initials is the quiescent center.

In recent studies with cultured Convolvulus roots (Phillips and Torrey[18]) it has been pointed out that within the quiescent center itself there exists a gradient of cells with differing cycle times. From experiments feeding ^{3}H-thymidine for various periods of time in culture and then preparing autoradiographs of sectioned roots, it is clear that the outer cells of the quiescent center, while showing prolonged cycle times compared to adjacent

TABLE I. DATA ON CELL CYCLE TIMES IN ROOT TIPS ‡

Plant	Type of Root	RCI	S	C	S & C	QC	Method	Reference
Zea mays	SR	12	25	48			CL	9
	(2 cm)	12	28	55		239	MA	9
Zea mays	(7 da) PR	14	22				PL	11
Zea mays		14	15-24			370	PL	3
Sinapis alba	SR	35			50		CL	10
	(1-2 cm)	25			50		MA	10
		35			32	520	MA	10
Vicia faba	SR (6 da)	44	37			292	MA	13
	SR (17 da)	53	27			228	MA	13
Allium sativum	PR	28	34				CL	26
	(3 da)	27	26			165	PL	26
	(1-2 cm)	33	35			173	MA	26
Convolvulus arvensis	CR	13	21	27		420	MA	19

‡ SR = seedling root; PR = primary root; CR = cultured root; RCI = root cap initials; S = stelar initials; C = cortical initials; QC = quiescent center; CL = continuous labelling; MA = metaphase accumulation; PL = pulse labelling.

initial cells, incorporate labelled DNA precursors before cells more central. The most central cells of the quiescent center therefore may be inferred to have the longest cycle times, i.e., the greatest duration of G_1. Thus one must conclude that the quiescent center cell population is not homogeneous but shows an approximately radial gradient of decreasing cell cycle times outward from the center.

The quiescent center is characterized at present largely in negative terms. Relative to the cell populations around them, cells of the quiescent center are metabolically sluggish, they synthesize DNA at infrequent intervals, they show low rates of amino acid incorporation (Clowes[8]), low rates of RNA synthesis judged from incorporation of labelled precursors (Clowes[6,7]) and presumably low respiratory activity judged from physiological studies of root segments (Jensen[14]) and the paucity of mitochondria seen in electron micrographs. Mitotic activity is as expected, very low, judged from mitotic counts (Jensen and Kavaljian[15]) or on less direct evidence from the studies reported above on DNA synthesis.

INTERPRETATIONS OF ROOT APICAL ORGANIZATION

How can one interpret or attempt to explain the existence of the quiescent center in the root apex? Several possibilities have suggested themselves. One idea is that the quiescent center results from a gradient in the oxygen supply with the central cells of the root apex most deficient in molecular oxygen and therefore inactive due to anaerobiosis. This theory has been explored by Amoore[2] who concluded that all cells in a root apex should normally not be limited by available oxygen supply under most conditions of development.

A second concept centers around the idea that the root apex, located as it is at the terminal end of an elongate heterotrophic structure, is always deficient or near-deficient in photosynthetic products from the shoot. Thus, one could envision the terminal region subject to the influence of limiting supplies, provided by specific pathways of vascular tissues, of carbohydrate such as sucrose or even a vitamin such as thiamine or nicotinic acid normally produced in the shoot and moved to the root. The importance of translocated photosynthate from shoot to root tip has been demonstrated dramatically in a number of seedling systems (e.g. Richardson[21]) and the remarkable effect of thiamine deficiency on root apical activity was shown by Addicott[1] among others. Yet under these conditions of deficiency, organizational patterns of the root apex were not simply modified; rather, the organization disappeared completely as all mitotic activity ceased. Also, Clowes[12a] has shown that ^{3}H-thymidine fed from labelled seed endosperm is, in fact, not differentially distributed around the quiescent center when thus provided from the proximal end of the root, but moves right across this region into the root cap. So it would seem unlikely that a specific distributional supply pattern of necessary substances and factors for root meristem activity alone controls the pattern of cellular relationships formed at the apex.

Another way of interpreting the organization of the root apex is to consider the quiescent center a site of specialized biosynthesis of a hormone essential to cell division in the root. At the center itself, a supraoptimal concentration for cell division occurs in the cells synthesizing the hormone. Within the cell population of the quiescent center, a gradient exists with the central-most cells "shut down" more completely than cells on the periphery. A very slight concentration drop to the critical concentration for cell division occurring along an outward diffusional gradient, would allow cells all around the quiescent center to divide at a maximal rate. These form the cells of the root apical initials whose distribution may be constrained also by other gradients determined perhaps by states of cellular differentiation within the different cell populations. Progressively more distant from the quiescent center, one would find a decreasing concentration of the hormone reflected in a transition from the rapidly dividing apical initials through the still dividing primary meristematic tissues and thence into the non-dividing cells of the mature primary tissues of the root. Such a sustained pattern of apical organization would depend upon a continuous synthesis at the source and standing wave concentrations of a hormonal stimulus radiating from the source as envisioned by Turing[28] and discussed by Wardlaw[29] in reference to the shoot apex. Removal of the hormone from the system could be by transport out of the apex or by combination with components of reactive cells which would "use up" the hormone in the processes of development.

On present evidence, one cannot decide what hormone best fits this hypothetical role of root organizer substance. Auxin, gibberellin, or cytokinin could qualify in that each has been demonstrated to occur in the apex in relatively high concentration. Each can be demonstrated in roots to become inhibitory at supraoptimal concentrations. Gibberellins and cytokinins are reported to move away from the root apex while auxin appears to be transported toward the apex. In no case do we have conclusive evidence concerning new synthesis in the apex itself of any of these known hormones.

Let us make the assumption, without documenting the case here, that the root apex is a site of cytokinin biosynthesis (cf. Weiss and Vaadia[30], Short and Torrey[24]). It is less justifiable from the available evidence but none the less attractive to make the hypothesis that auxin is largely shoot-produced (perhaps in shoot apices and young leaves) and that the auxin reaching the root apex is due largely to its translocation in a polar fashion from the proximal to the distal regions of the root (Bonnett and Torrey[4], Kirk and Jacobs[16], Scott and Wilkins[22]). The further assumption can be made that all cell-division activity is dependent upon available, non-limiting supplies of auxin interacting with other hormones (Skoog and Miller[25], Patau, Das and Skoog[17], Setterfield[23]). And, of course, one must further assume an available supply at the root apex of shoot-originating carbohydrates such as sucrose and vitamins such as thiamine, which are all essential ingredients for continued meristematic activity at the root apex.

In a well-established, rapidly growing root, the quiescent center is clearly definable and involves a large number of cells--of the order of a thousand or more in a root the size of Zea or Convolvulus. From autoradiographic evidence described above, one can conclude that there exists in such a cell population a gradient with respect to the tendency to divide or to complete the cell cycle. The internal-most cells seem to have the longest cycle times and this time is progressively shorter until one reaches the periphery of the quiescent zone where an abrupt discontinuity exists--where cells are in active division--i.e., the apical initials themselves. One can visualize this gradient related to a chemical gradient in cytokinin formation--supraoptimal in the center, optimal in the regions of the initials, then limiting both distally and proximally. At present, this hypothesis is quite unproved.

This view of the organization of the root apex leads to a number of further interesting ideas about root development. In seedling embryo radicles and in initiating lateral roots, Clowes showed by autoradiography that the initiating organ first lacks a quiescent center and then develops a progressively enlarging population of cells with low rates of DNA synthesis. According to our hypothesis, such initiation is the reflection of an activation of hormone synthesis in the cells of the emerging quiescent center. Such activation may occur first as a result of an unequal cell division such that one of the progeny perpetuates a special biosynthetic activity which at first leads to localized cell proliferation and then to a shutting down of the population formed. Such a view fits well with experimental evidence (Torrey[27]) showing the subtle influence on lateral root initiation of slight changes in externally supplied cytokinin concentrations--varying from marked stimulation to complete inhibition. This view also leads to further experiments subject to test. One could follow hormone synthesis during lateral root initiation using extraction and bioassay. Such an experiment might also be done on callus tissue induced to initiate root primordia by hormone manipulation.

The activation of specialized biosynthetic pathways is not a novel idea and must be involved in most cytodifferentiation. Demonstration of such activation with respect to hormone synthesis is most clearly documented in the case of crown-gall transformation (Braun[5]) in a phenomenon leading to uncontrolled cell proliferation. Here apparently several biosynthetic pathways are activated together and a disorganized meristematic mass results. Through rigidly controlled specific biosynthetic activations of analogous type, the developing plant could redirect tissue activities into newly organized systems. Perhaps bud initiation can be thought of in exactly comparable terms but involving another specific hormone synthesis--perhaps of auxin. Root apices as primary but not exclusive sites of cytokinin biosynthesis and shoot apices as primary but not exclusive sites of auxin synthesis would offer a simple model for polarized development in the plant. Earliest polarization in the developing embryo would then involve the cellular separation and elaboration of such controlling biosynthesis in the cell populations comprising the respective apices. How the plant initiates and controls specific biosynthetic activations is the fundamental problem of cytodifferentiation; how these activities are contained and sustained may well be the basic problem of organogenesis.

REFERENCES

1. Addicott, F. T. (1939) Vitamin B_1 in relation to meristematic activity of isolated pea roots. Bot. Gaz. 100: 836-843.

2. Amoore, J. E. (1961) Dependence of mitosis and respiration in roots upon oxygen tension. Proc. Roy. Soc., B 154:109-124.

3. Barlow, P. W. (1969) Phases of the cell cycle in the root meristem of *Zea mays*. in *Root Growth*. W. J. Whittington (ed.), Plenum Press, N.Y.

4. Bonnett, H. T., Jr. and Torrey, J. G. (1965) Auxin transport in *Convolvulus* roots cultured in vitro. Plant Physiol. 40: 813-818.

5. Braun, A. C. (1958) A physiological basis for autonomous growth of the crown-gall tumor cell. Proc. Nat. Acad. Sci. 44:344-349.

6. Clowes, F. A. L. (1956a) Localization of nucleic acid synthesis in root meristems. J. Exptl. Bot. 7:307-312.

7. ______________.(1956b) Nucleic acids in root apical meristems of *Zea*. New Phytol. 55:29-35.

8. ______________.(1958) Protein synthesis in root meristems. J. Exptl. Bot. 9:229-238.

9. ______________.(1961) Duration of the mitotic cycle in a meristem. J. Exptl. Bot. 12:283-293.

10. ________________. (1962) Rates of mitosis in a partially synchronous meristem. New Phytol. 61:111-118.

11. ________________. (1965) The duration of the G_1 phase of the mitotic cycle and its relation to radiosensitivity. New Phytol. 64:355-359.

12. ________________. (1968) The DNA content of the cells of the quiescent center and root cap of *Zea mays*. New Phytol. 67: 631-639.

12a. ________________. (1970) Nutrition and the quiescent centre of root meristems. Planta (Berl.) 90:340-348.

13. ________________. and Hall, E. J. (1962) The quiescent centre in root meristems of *Vicia faba* and its behaviour after acute X-irradiation and chronic gamma irradiation. Rad. Bot. 3:45-53.

14. Jensen, W. A. (1955) A morphological and biochemical analysis of the early phases of cellular growth in the root tip of *Vicia faba*. Exptl. Cell. Res. 8:506-522.

15. Jensen, W. A. and Kavaljian, L. J. (1958) An analysis of cell morphology and the periodicity of division in the root tip of *Allium cepa*. Amer. J. Bot. 45:365-372.

16. Kirk, S. C. and Jacobs, W. P. (1968) The movement of 3-indoleacetic acid-^{14}C in roots of Lens and Phaseolus. pp. 1077-1094 in Biochemistry and physiology of plant growth substances (Ed. F. Wightman and G. Setterfield). Runge Press, Ottawa.

17. Patau, K., Das, N. K. and Skoog, F. (1957) Induction of DNA synthesis by kinetin and indoleacetic acid in excised tobacco pith tissue. Physiol. Plant 10:949-966.

18. Phillips, H. L., Jr. and Torrey, J. G. (1971a) The quiescent center in cultured roots of *Convolvulus arvensis* L. Amer. J. Bot. (in press).

19. ________________. and ____________. (1971b) Duration of the cell cycle in cultured roots of *Convolvulus*. Amer. J. Bot. (in press).

20. ________________. and ____________. (1971c) DNA synthesis in root cap cells of cultured roots of *Convolvulus*. Plant Physiol. (in press).

21. Richardson, S. D. (1953) Studies of root growth in Acer saccharinum L. I. The relation between root growth and photosynthesis. Proc. Koninkl. Nederl. Akad. Weten. Amsterdam C. 56:185-193.

22. Scott, T. K. and Wilkins, M. B. (1968) Auxin transport in roots. II. Polar flux of IAA in Zea roots. Planta (Berl.) 83:323-334.

23. Setterfield, G. (1963) Growth regulation in excised slices of Jerusalem artichoke tuber tissue. Symp. Soc. Exp. Biol. 17:98-126.

24. Short, K. C. and Torrey, J. G. (1971) Cytokinins in seedling roots of pea. Plant Physiol. (in press).

25. Skoog, F. and Miller, C. O. (1957) Chemical regulation of growth and organ formation in plant tissues cultured in vitro. Symp. Soc. Exp. Biol. 11:118-131.

26. Thompson, J. and Clowes, F. A. L. (1968) The quiescent centre and rates of mitosis in the root meristem of Allium sativum. Ann. Bot. 32:1-14.

27. Torrey, J. G. (1962) Auxin and purine interactions in lateral root initiation in isolated pea root segments. Physiol. Plant 15:177-185.

28. Turing, A. M. (1952) The chemical basis of morphogenesis. Phil. Trans. Roy. Soc. B. 237:37-72.

29. Wardlaw, C. W. (1953) A commentary on Turing's diffusion-reaction theory of morphogenesis. New Phytol. 52:40-47.

30. Weiss, C. and Vaadia, Y. (1965) Kinetin-like activity in root apices of sunflower plants. Life Sci. 4:1323-1326.

* The author acknowledges the research work and helpful discussions of his graduate students and post-doctoral fellows but takes sole responsibility for the unsupported speculations included herein.

-DISCUSSION-

CONGER, A. - Is it possible that the reduction in size of the quiescent center is a response to the accumulated radiation which has inactivated an appreciable number of cells?

CLOWES - It's partly that, but there is a gradation across the quiescent center, with maximum quiescence in the center, and an increase in the rate of mitosis at the edge.

CONGER, A. - Radiation cytologists have shown that long exposures to thymidine damage cells. Perhaps radiation is merely killing some of them at the quiescent center, resulting in a reduction in its size. Would this vitiate the gradient argument for the quiescent center?

LANGE - It's the cells on the periphery of the quiescent center which one would expect the thymidine to damage. On this basis then, the inner cells should become the most mitotically active.

VAN'T HOF - Is the quiescent center in dynamic equilibrium with its neighbors?

TORREY - It really must be in dynamic equilibrium in order to sustain an ongoing population of cells. "Quiescence" is simply an operational definition. Everything we know about it (the quiescent center) is negative. It is a region which does not incorporate ^{3}H TdR.

VAN'T HOF - But it changes in size with time and you postulate that there are different cycle times within the quiescent center itself. Where does the quiescent center begin and the other population end?

TORREY - In a rapidly elongating root the size of the quiescent center is more or less constant. My interpretation is that this maintenance of size reflects differences in cycle time.

VAN'T HOF - But with a 2 1/2-day exposure to label it diminishes in size. Is this a tracer effect?

CLOWES - When you apply external radiation the cells that are most predominantly stimulated into mitosis are those of the center of the quiescent center. If there was any effect of radiation from the soluble ^{3}H TdR it would tend to stimulate these cells. Whereas Dr. Torrey observes an effect at the outside so this is quite different from a "pure" radiation effect.

VAN'T HOF - If all the cells of the quiescent center are labeled after 120 hours, how can the cycle time be 420 hours?

TORREY - This calculation is based upon the metaphase accumulation data. Incorporation may involve only a section of the cycle time.

SAVAGE - There are two potential effects from ^{3}H TdR in these long treatments, (a) the internal radiation and (b) the uptake of and prolonged exposure to the chemical itself.

HABER - Will labeled cells of the quiescent center move out?

CLOWES - Yes, they will.

SMITH - The universal occurrence of a quiescent center implies that it may have an evolutionary importance. In order to have a biochemical center does one have to have cells that are in "resting" stages?

TORREY - I don't think it's necessary to shut down other operations to do this.

SMITH - Can interphase cells contribute hormones to dividing cells?

TORREY - At present we don't know.

DAVIDSON - Do the metaphase accumulation curves include restitution? If they do, then the estimate of cell cycle duration will be incorrect.

TORREY - No. The restitution nuclei were included as part of the total population but the accumulation curve never included a chromosome set that was going back into interphase.

SMITH - Only a very small proportion of cytokinin is associated with t-RNA.

TORREY - There is considerable concern among cytokinin experts about whether the free or the bound cytokinin (t-RNA) is the important one. The overwhelming amount of free cytokinin is in the tip of the root. Whether or not it's the result of a synthetic (building) or a degradation process is not yet known.

SMITH - But the bound fraction may be the important one.

TORREY - Except that externally applied cytokinin appears not to be bound.

VAN'T HOF - With respect to the results obtained in your system Dr. Torrey, it depends on what kind of cell is treated with cytokinin. If one treats the pea root meristem with the same concentration as you have used for your segments, mitotic cycle activity is inhibited, whereas in segments it's enhanced.

SAVAGE - We regularly obtain cytoplasmic labeling in pollen and have ruled out the possibility of a decomposed-thymidine effect by utilizing very fresh thymidine. Could the cytoplasmic labeling be mitochondrial labeling?

VAN'T HOF - Stationary phase meristems (i.e., with no cells in "S") fed tritiated thymidine show cytoplasmic label but its buoyant density is identical to that of labeled nuclear DNA.

CONGER, A. - So it is not mitochondrial DNA?

VAN'T HOF - "Alaska" cytoplasmic DNA has a buoyant density near to that of nuclear DNA, and can be labeled with tritiated thymidine.

TORREY - Proplastids too can be labeled and they are certainly present.

DAVIDSON - You stated that cycle times differed in G_1?

TORREY - It is difficult to determine for individual parts of the meristem. We have no direct evidence on this. We have depended heavily on Clowes' arguments that this is where the effect is.

CLOWES - From pulse label studies it appears that the cells of the quiescent center do have a much longer G_1.

LANGE - Is the G_1 cell of the quiescent center proceeding through a very slow set of syntheses or is it sitting there waiting to be told "it is now time to prepare to divide" -- and then it starts doing things at the normal rate? Both of those cases are G_1 by the subtraction method $T_c-(S+G_2+M)$ but the cells are doing quite different things. Thus it may be more useful to use the concept of the G_0 cell in such situations.

CLOWES - I think these can be distinguished if you measure the growth fraction, and if you measure G_1 by subtraction of all the other bits of the cycle. For those cells which are actually cycling one can determine a G_1. It cannot be measured in non-cycling cells.

MITOTIC CYCLE REGULATION IN THE MERISTEM OF CULTURED ROOTS: THE PRINCIPAL CONTROL POINT HYPOTHESIS*

J. VAN'T HOF and C. J. KOVACS[1]

Biology Department, Brookhaven National Laboratory

Upton, L.I., New York 11973

INTRODUCTION

This paper was deliberately limited to the concept of and experimental evidence for two principal control points in the mitotic cycle of root meristem cells. The root meristem has been the vehicle for much research in cell population kinetics. The work of Laughlin[9,10], with *Allium* in 1919 and Brown[1] with *Pisum* in 1951 on temperature effects on the mitotic cycle attest to an early recognition of its advantages for such studies. Even today few experiments match the precision, breadth, and insight of Gray and Scholes[2] in their growth and cellular analysis of *Vicia faba* roots. The work of Gray and Scholes and that of Lea[11] undoubtedly contributed to the formulation of the universally accepted Howard and Pelc[4] model of the mitotic cycle. The model coupled with the tools of radioisotope and high resolution autoradiography, also developed with *Vicia*[14], and the theory of Quastler[12] and Quastler and Sherman[13] formed the foundation of what is currently known as cell population kinetics. Wimber[26] was one of the first to use these tools to describe the mitotic cycle of *Tradescantia* root tip cells and since the first experiments, much information has been gathered about the cell division cycle in plants. For the most part, however, the information is descriptive and lacks useful notions of how the mitotic cycle is

*Research supported in part by the U.S. Atomic Energy Commission and in part by the U.C. Public Health Service (Fellowship 1F02 CA-42865-01 from the National Cancer Institute to C.J.K.).
[1]Present address: Division of Dermatology, Hahnemann Medical College, Philadelphia, Pa. 19102.

regulated. For this reason it is our intention to put forth a concept embodied in the two principal control point hypothesis that is at once useful, testable, and hopefully will increase our understanding of the mitotic cycle in plant root meristems.

The paper is divided into five parts. First there is a glossary which may be unnecessary for those familar with the work on stationary phase meristems but should be useful for those who need a quick definition of a particular symbol when it is not explained in the text. Second, there is the statement of the principal control point hypothesis. The third part consists of results obtained with *Pisum* roots which serve both as an introduction to the kind of data produced from starvation experiments and as a standard to which results obtained with other species may be compared. Part four deals with the functional independence of the principal control points as revealed by ionizing radiation studies. The discussion centers on cells in G1 which are different biochemically and which correspondingly express radiation damage at different times in the mitotic cycle. The fifth and final part returns to the simpler carbohydrate starvation experiment as performed with *Vicia* and *Helianthus* root tips and considers variation of principal control point expression in different species.

GLOSSARY

S	--	Period of DNA synthesis
Ts	--	Duration of S
Tds	--	Delay of entry of S.P. G1 cells into S
M	--	Period of mitosis
Tm	--	Duration of M
Tdm	--	Delay of entry of S.P. G2 cells into M
G2	--	Postsynthetic premitotic period
Tg2	--	Duration of G2
G1	--	Postmitotic presynthetic period
S.P.	--	Stationary phase
^{3}H-TdR	--	Tritiated thymidine
P.C.P.	--	Principal Control Points or Point depending on context.
R	--	Rate of entrance of cells into S (% labeled cells per hr)

THE PRINCIPAL CONTROL POINT HYPOTHESIS

Cellular proliferation in meristematic tissue of cultured pea roots is the result of sequential processes which are recognized cytologically as the mitotic cycle. When environmental conditions prevent cells from satisfying these regulating processes, cycle progression eventually ceases. For instance, it is known that continuous cell proliferation in the meristem of cultured pea roots requires a high level of metabolism and if starved of carbohydrate a stationary condition is produced with cells arrested in G1 and G2[16,17,24]. Such meristems are defined as stationary phase (S.P.) meristems. That cells stop preferentially in G1 and G2 after prolonged carbohydrate starvation inferred the pre-DNA synthetic and pre-mitotic periods possess a quality not shared by either S or M. Consequently, Van't Hof[21] proposed that in pea root meristems, the mitotic cycle is governed by two Principal Control Points (P.C.P.); one in G1 that regulates the G1→S transition and another in G2 that controls the G2→M transition.

The points or loci at which the cells arrest are considered metabolic blocks that prevent the initiation of DNA synthesis and mitosis and which are relieved when exogenous carbohydrate is replenished. Resumption of cell progression in the cycle is not immediate, however, as certain preparations are necessary before entry into S or M. It is during the interval between the provision of carbohydrate and the first appearance of cells in S and M that metabolic and molecular requirements for the G1→S and G2→M transitions are met[6,7,19,20,23,24]. The delay of cell entry from G1 to S (T_{ds}) or from G2 to M (T_{dm}) are cytological parameters that measure the ability of cells to complete the requirements for mitotic cycle advancement or stated in another way, T_{ds} and T_{dm} measure the length of time needed by former S.P. G1 and G2 cells to complete the requirements of the respective P.C.P. To date, experiments with S.P. meristems have shown that the G1 and G2 P.C.P. have requirements that are not met: a) without carbohydrate[15,16,18], b) with carbohydrate but without oxygen[20,23], c) in the presence of an uncoupler of oxidative phosphorylation[23], and d) in the presence of inhibitors of protein synthesis[22,23]. From these observations it was concluded that satisfaction of the P.C.P. requirements has an immediate need for proteins the synthesis of which is energy dependent.

RESOLUTION OF PRINCIPAL CONTROL POINTS IN PISUM

Pioneer experiments performed with Pisum root meristems demonstrated that cell progression in the cycle required carbohydrate and that the absence of sugar resolved the P.C.P.[15,16]. The results of one such experiment are shown in Fig. 1 and their consideration serve a three-fold purpose. First, they introduce the reader to the kind of data gathered, second, they acquaint the reader with the terminology, and third, the data will be used as a standard to which those obtained with Vicia faba and Helianthus annuus are compared. The results shown in Fig. 1 were obtained with 72 hr S.P. root tips. It will be recalled that during these 72 hr the cells arrest in G1 and G2 and, therefore, at zero hr when 2% sucrose and ^{3}H-TdR were given no labeled interphase cells were observed (Fig. 1a). It took 11 hr before the first labeled cells were seen (T_{ds}) after which time they entered S at a rate of 2% per hr (R). These same cells viewed as labeled mitotic figures began to divide at 17-18 hours (Fig. 1b) a period of time representing the sum of T_{ds}, T_s and T_{g2}. Preceding the former S.P. G1 cells to M were those that initiated cycle progression from G2. The former S.P. G2 cells, of course, were unlabeled and none divided until a delay of 11 hours had passed (Fig. 1c) demonstrating a T_{dm} of 11 hours. Both cells in G1 and G2 during the S.P. required 11 hr in the presence of

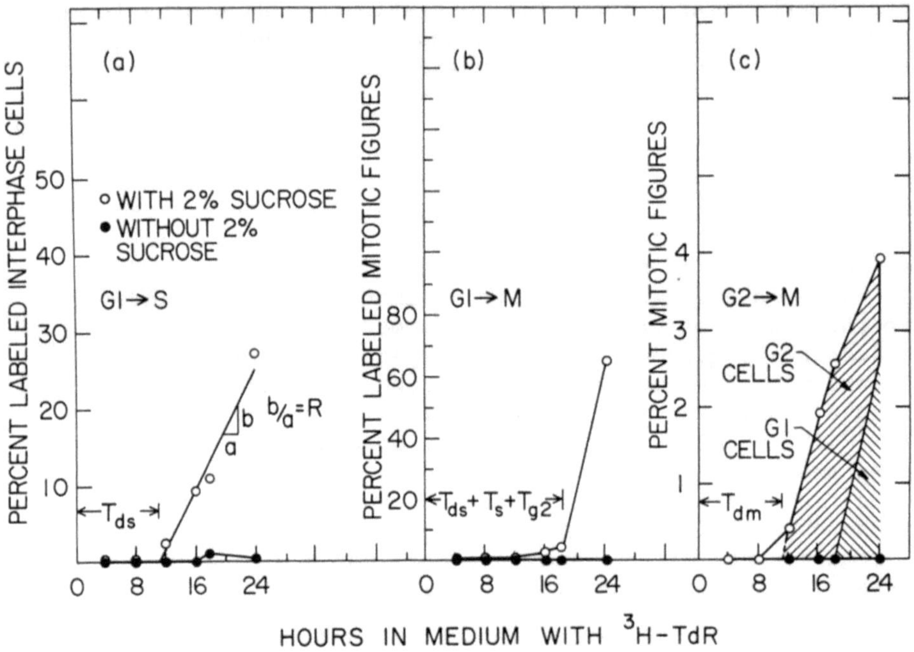

Fig. 1. The resumption of progression in the mitotic cycle by cells in primary root tips of Pisum sativum. Root tips were cultured for 72 hr in White's medium[25], pH4.7, without sucrose. At time zero they were transferred to medium that contained either 2% sucrose and ^{3}H-TdR or only ^{3}H-TdR. See refs. 15 and 16 for details. a) The entrance of cells into S from G1 expressed as the increase in labeled interphase cells as a function of time. b) The entrance of former G1 cells into M expressed in terms of percent labeled mitotic figures. c) The entrance of cells into M, first from G2 and then G1, expressed as the change in percent total mitotic figures with time.

carbohydrate before they were able to advance to S and M respectively and during the 11 hr delay they were synthesizing the necessary requirements for the respective P.C.P. In contrast, if not supplied carbohydrate, no cell advancement took place a result indicative of dependency on exogenous carbohydrate for the reversal of the S.P. Further, the fact that cells arrested in G2 remained there for 72 hr and when provided sucrose did not re-initiate DNA synthesis demonstrated that the G2 P.C.P. was unidirectional oriented toward mitosis. Similarly the linear entrance of cells from G1→S also implied that in G1 the cells remained in position relative to each other during the S.P. If they had either slowly moved forward to or backward from the G1/S boundary a linear entrance would be unlikely.

FUNCTIONAL INDEPENDENCE OF THE P.C.P.

Experiments with S. P. meristems demonstrated that cells which previously divided asynchronously were preferentially arrested in G1 and G2. Additional work showed that S.P. cells in G1 which were induced to enter S and then further subjected to carbohydrate starvation stopped in G2[20,21]. Thus, the Pisum root tip does not consist of two cell populations, one that exclusively stops in G1 and another that only stops in G2 as proposed by Gelfant[3] for other tissues. That the Pisum root meristematic cells arrest in either G1 or G2 hints that the P.C.P. are functionally independent. Further evidence for functional independence was obtained from irradiation experiments involving two different populations of G1 cells[22]. The first population was that of 48 hr S.P. root meristems which can not advance to S in the presence of inhibitors of protein synthesis when provided sucrose. The other population of G1 cells was obtained by synchronization procedures that result in an accumulation of cells in the meristem at the G1/S boundary[5,6]. The synchronized G1/S cells can not be prevented from proceeding to S even though protein synthesis inhibitors are present in the medium[7,22]. In terms of the P.C.P. hypothesis the S.P. population can not advance to S when protein synthesis is impaired because the requirements of the P.C.P. in G1 can not be met. In contrast, the synchronized G1/S population was not affected by the same inhibitors because during the synchronization procedures the G1 P.C.P. requirements were completed[7,8]. Keeping in mind the biochemical differences between these two populations of G1 cells, results obtained with ionizing radiation will be discussed. In Fig. 2 are shown data obtained from G1 cells in 48 hr S.P. meristems. On the left (Fig. 2a), is shown the manner in which cells entered S. In the control a 4 hr T_{ds} was observed and the cells entered S at approximately 2% per hr. If irradiated the G1 cells required not 4 but 10 hr to enter S. The additional 6 hr was ascribed to recovery time needed to remove damage caused by irradiation[17,20,22]. On the right (Fig. 2b) are shown curves that describe the entrance of former G1 cells into M. The first cells arrived at M at 12 hr in the control and at 18 hr in the irradiated sample, the difference being 6 hr and equivalent to the difference in T_{ds}. The T_s+T_g2 of each population was the same and the delayed arrival at M of the irradiated cells was mostly due to an extended T_{ds}.

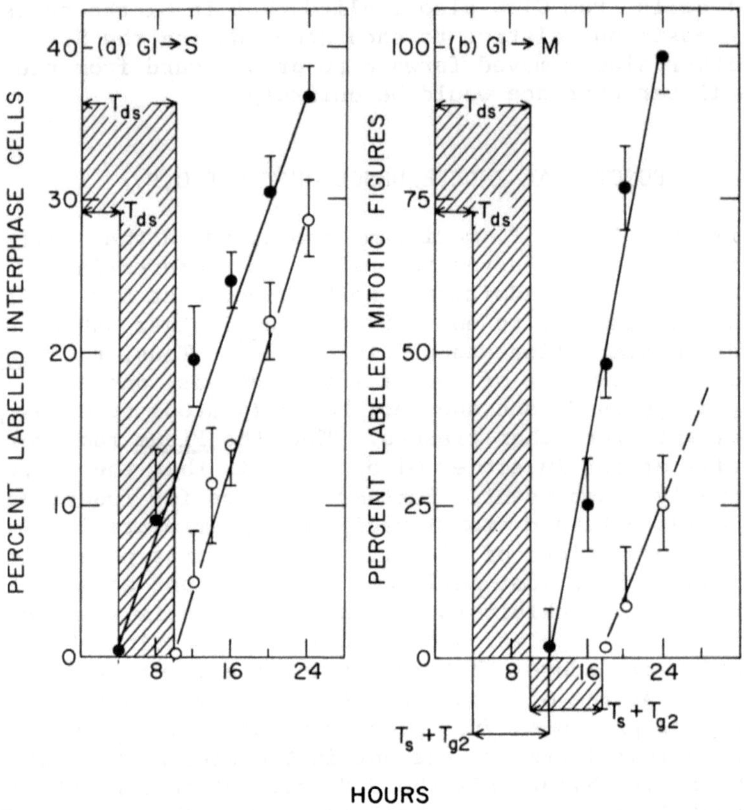

Fig. 2. The resumption of progression in the mitotic cycle by cells in primary root tips of Pisum. Root tips were cultured for 48 hr in White's medium[25], pH 4.7, without 2% sucrose. At zero hr some were exposed to 300 R of γ-rays, 300 R/min, and then transferred to medium with 2% sucrose and ^{3}H-TdR. a) The entrance of cells from G1 into S expressed as the increase in labeled interphase cells. Solid circles, control; open circles, irradiated. b) The appearance of former G1 cells in mitosis expressed as an increase in number of labeled mitotic figures.

In Fig. 3 are shown the results obtained with cells synchronized at the G1/S boundary. At time zero those represented by the open circles were irradiated with 300 R of γ-rays and sucrose was provided to release the cells into S. After release at the intervals shown, the meristems were pulse labeled with ^{3}H-TdR to determine the percent of cells in S and to trace the passage of cells through S. The curves obtained for the control and irradiated populations were almost identical indicating that the γ-rays neither stopped the cells from going to S nor caused them to have a prolonged S. Following the cells as they continued through the cycle to M revealed that the control cells began to divide at 9 hr and the irradiated cells at 13 hr. Since T_s was unchanged and there was no delay of entry into S, the irradiated

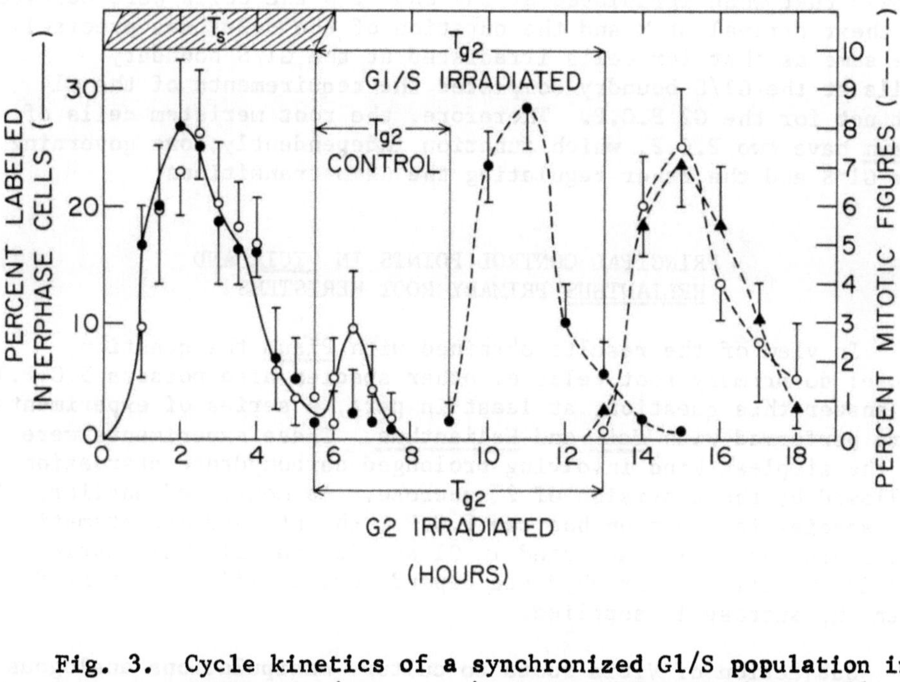

Fig. 3. Cycle kinetics of a synchronized G1/S population in the root tip of <u>Pisum</u>. (________) percent of interphase cells synthesizing DNA after release into the S stage of the cycle; (-------), appearance of a G1/S population at mitosis. Solid circles, control G1/S population; open circles, G1/S population irradiated at G1/S; solid triangles, G1/S population irradiated at the end of S, 5.5 hr on the graph. For labeled interphase cells, each data point is that fraction of cells labeled after a 30 min pulse with ^{3}H-TdR.

cells must have stopped temporarily in G2. Therefore, unlike irradiated S.P. G1 cells which were delayed in G1, the synchronized G1/S cells were delayed in G2. In terms of the P.C.P. hypothesis these results suggest: a) that in S.P. G1 cells both P.C.P. were damaged by irradiation and were recovered simultaneously during the extended T_{ds} (Fig. 2a), b) that in synchronized G1/S cells both P.C.P. were also damaged but no delayed entry into S occurred because during the synchronization procedures the G1 P.C.P. was completed, and c) that the absence of a delayed entry into S provided no time for the G2 P.C.P. to recover and when the synchronized cells arrived at G2 they waited until such recovery was finished before proceeding to M. In support of this last point, data are included in Fig. 3 from an experiment in which synchronized cells were irradiated at the end of S (at 5.5 hr on the graph) and the arrival time at M determined. The results showed that when irradiated at the end of S the cells were delayed in their arrival at M and the duration of the delay was precisely the same as that for cells irradiated at the G1/S boundary. Cells at the G1/S boundry completed the requirements of the G1 but not for the G2 P.C.P. Therefore, the root meristem cells of *Pisum* have two P.C.P. which function independently; one governing the G1→S and the other regulating the G2→M transitions.

PRINCIPAL CONTROL POINTS IN *VICIA* AND *HELIANTHUS* PRIMARY ROOT MERISTEMS

In view of the results obtained with *Pisum* the question arose: do primary root cells of other species also possess P.C.P.? To answer this question, at least in part, a series of experiments were performed with *Vicia* and *Helianthus*. These experiments were of the simplest kind involving prolonged carbohydrate starvation followed by the provision of 2% sucrose. As mentioned earlier, if the species in question has two P.C.P., the starved meristematic cells are eventually arrested in G1 and G2 and, if this stationary condition reflects carbohydrate dependency, it will be reversed when the sucrose is supplied.

Subjection of *Vicia* roots to culture manipulations analogous to those used with *Pisum* produced the results shown in Fig. 4. *Vicia* meristems, like those of *Pisum* had cells neither in S nor in M after 72 hours of starvation and even an additional 28 hours in fresh medium without sucrose failed to induce DNA synthesis (Fig. 4a) or mitosis (Fig. 4c). Only when provided carbohydrate did the S.P. cells eventually enter S or begin to divide after exhibiting a T_{ds} of 7-8 and a T_{dm} of 7 hours. Thus *Vicia* cells have two P.C.P. but they are effected differently than those of *Pisum*. This difference was detected in R which was 2% for *Pisum* (Fig. 1a) and 1% per hour for *Vicia* cells (Fig. 4a). Considered in its simplest form, R should vary directly with the relative proportion of S.P. G1 cells, i.e., the more cells stopped in G1, the higher R would be. Assuming all other factors equivalent, Vicia has proportionally fewer cells in G1 than S.P. meristems of Pisum. Indeed if this were the case, then the converse should also apply, that is, the number of former S.P. G2 cells that divided after sucrose replenishment should be greater in Vicia than in Pisum. Examination of the data, presented on the same scale, in Figs. 1c and 4c offer verification of this expectation. The number of unlabeled mitoses observed in Vicia (Fig. 4c) always exceeded that of Pisum (Fig. 1c) in these and in other experiments where measurements continued up to 36 hours.

Helianthus root meristem cells, in contrast to those of Pisum and Vicia, were adversely effected by 72 hours of starvation so the duration of deprivation was reduced to 48 hours. Nevertheless, the data shown in Fig. 5 include some informative results obtained with 72 hour S.P. cells that will be discussed later. The meristems of Helianthus proved unique in another way for prolonged

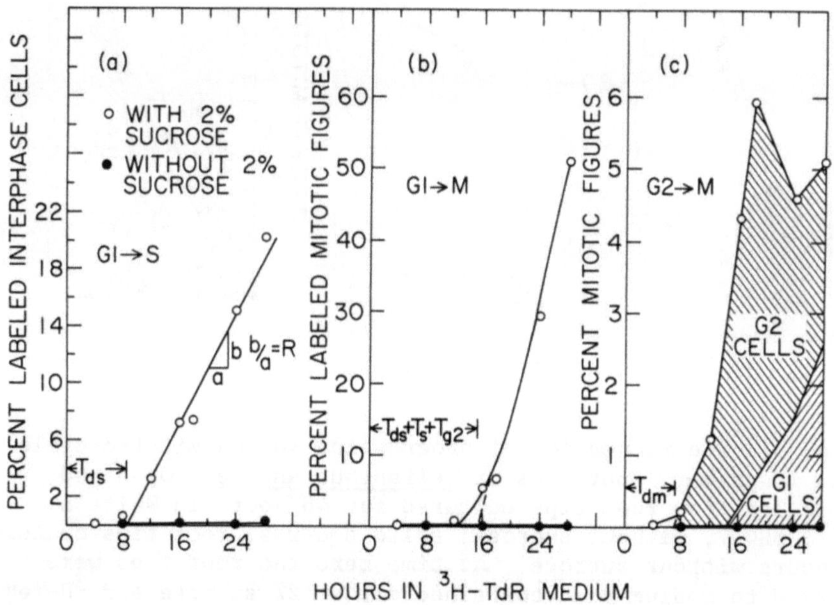

Fig. 4. The resumption of progression in the mitotic cycle by cells in primary root tips of Vicia faba. Conditions of the experiment and graphical representations are as indicated in the legend of Fig. 1 except White's medium was pH7.0 adjusted with NaOH.

starvation never produced a true S.P. Whether starved for 48 or 72 hours there were always a few cells in S and M and the percentage was reduced with increased deprivation. Hence, certain cells "leak" from G1 to S (Fig. 5a) and from G2 to M (Fig. 5c). The "leakage" rate into S in 48 hour S.P. meristems was approximately 0.8% per hour and that for 72 hours S.P. meristems was nearly 0.3% per hour (Fig. 5a). The majority of cells, however, became stationary and most of these stopped in G1. When replenished with sucrose they entered S at a rate (R) of 8% per hour (Fig. 5a) and those few that stopped in G2 proceeded to M (Fig. 5c). The "leaky" cells had a T_{ds} and T_{dm} of zero and the cells that arrested in G2 apparently had a T_{dm} of zero also as no delay was detectable in

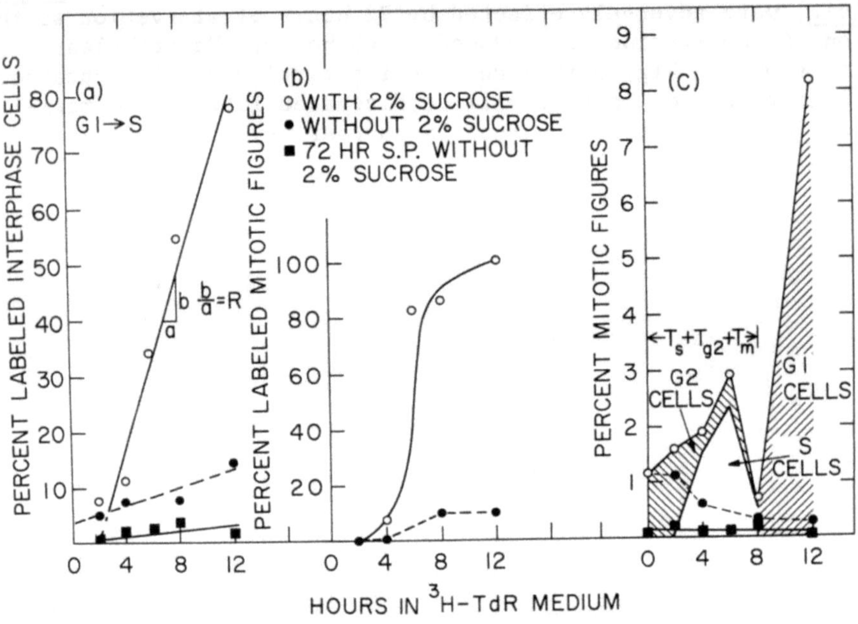

Fig. 5. The resumption of progression in the mitotic cycle by cells in primary root tips of Helianthus annuus. Open and solid circles from root tips cultured for 48 hours in White's medium[25], pH4.7, without sucrose; solid squares, root tips cultured for 72 hours without sucrose. At time zero the root tips were transferred to medium that contained either 2% sucrose and ^{3}H-TdR or only ^{3}H-TdR. a) The enterance of cells from G1 into S expressed as percent labeled interphase cells. b) The enterance of cells from S and G1 into M expressed as the increase in percent labeled mitotic figures. c) The enterance of cells into M, first from G2, then from S, and finally from G1 expressed as the change in percent total mitotic figures with time.

their entry to M when sucrose was finally supplied (Fig. 5c). Conversely, if arrested in G1, the S.P. cells required approximately 2.5 hours before proceeding into S. In this instance, T_{ds} was not determined by extrapolation of the labeled interphase cell curve to the abscissa, but was estimated from the point of interception between the percent labeled interphase cell curves traced for 48 hour S.P. meristems with and without sucrose (Fig. 5a). The latter, in this case, constituted a baseline established by the "leaky" cells while the former represented a deviation from this baseline induced by carbohydrate provision. Because the high R value indicated that most S.P. cells were in G1, it was expected that their arrival at M be marked by a very rapid rise in percent mitotic figures. The data, shown in Fig. 5c, verify this expectation. Beginning at 8 and extending to 12 hours the percentage rose sharply reflecting the division of cells formerly arrested in G1. Since they arrived at M at 8 hours after an initial delay of 2.5 hours, the T_s+T_g2 for these cells was 5.5 hours.

Helianthus root meristem cells differed from those of Pisum and Vicia in at least three ways: first, not all cells became stationary with carbohydrate starvation; second, of those arrested, most accumulated in G1; and third, a T_{ds} of 2.5 hours, a R of 8% per hour, and an apparent T_{dm} of zero suggested the presence of a predominent P.C.P. located in G1.

The results obtained with Pisum, Vicia, and Helianthus summarized diagrammatically in Fig. 6 provide a simple outline for the variation of P.C.P. expression as revealed by the distribution of cells in the mitotic cycle after starvation. Diagram I describes the cells in the cycle before starvation and characterizes the proliferative status of each species analyzed. Diagram II represents the distribution of Pisum root meristem cells after the S.P. was established. No cells were in S or M and slightly more cells accumulated in G1 than in G2. Diagram III depicts the distribution of Vicia root meristem cells after starvation. There were no cells in S or M and of those arrested, most were in G2. Finally, the bottom diagram (IV) represents the cell distribution in starved Helianthus root meristems. Under conditions where the duration of starvation was not deleterious, cells were present in each cycle period, although the major portion were arrested in G1.

IMPLICATIONS OF VARIED CONTROL POINT EXPRESSION

It is evident that carbohydrate starvation and replenishment experiments provide information about the mitotic cycle that was heretofore unobtainable. Though the P.C.P. of Pisum and Vicia cells displayed subtle and important differences it was felt that, at the moment, these were secondary to the more obvious differences of Helianthus cells the majority of which either lacked or possessed a very weak G2 P.C.P. and others that perhaps had neither the G1 nor the G2 P.C.P.

Implicit in the statement that the P.C.P. are resolved by carbohydrate starvation is the realization that cells failing to respond under these conditions to the G1 and G2 P.C.P. either lack the P.C.P. and possess other regulatory mechanisms or have alternative pathways for satisfying their requirements. The "leaky" cells in Helianthus meristems are examples of such deviates. While most cells stop in G1 and a few apparently arrest in G2, a small number continue to advance in the mitotic cycle in spite of a lack of carbohydrate. Analysis of the data in Figs. 5a and 5c for S.P. meristems not provided sucrose established these cells as possessing an unusual metabolism possibly involving autolytic processes. This conclusion was based on the fact that both labeled interphase cells (Fig. 5a) and mitoses (Fig. 5c) were fewer in meristems

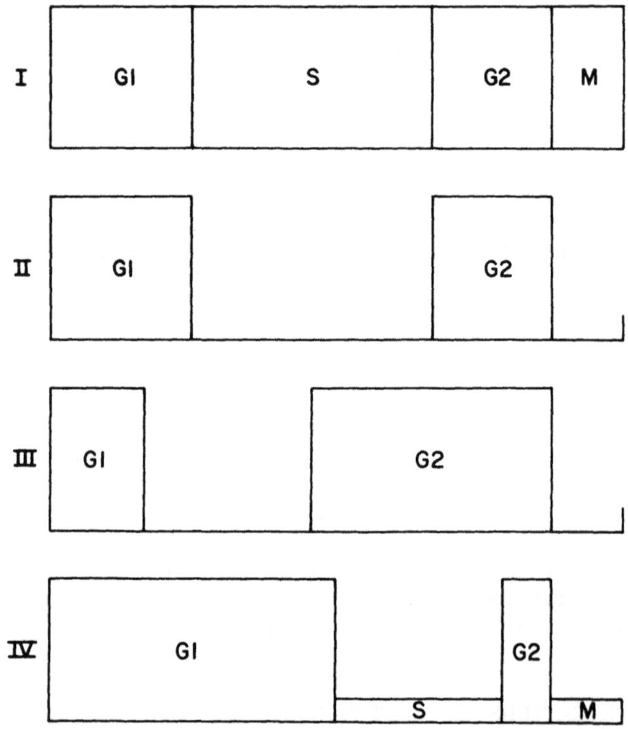

Fig. 6. Diagrams representative of cell distribution in the mitotic cycle of primary root meristems of Pisum, Vicia and Helianthus before carbohydrate starvation (I); in Pisum (II) and Vicia (III) after 72 hr of starvation, and in Helianthus after 48 hours of starvation (IV).

starved for 72 hours than in those starved for 48 hours. Had the "leaky" cells been so unique as to arrest in either S or M, the 72 hour S.P. roots should have had more, not less, labeled cells and mitoses. Further, if the "leaky" cells were capable of metabolism not deleterious to themselves, they would have maintained their numbers and been independent of the starvation period. That the number of "leaky" cells actually became less suggested persistent mitotic cycle activities were carried out at the expense of survival or, when finally arrested they too stopped in G1.

In summary the comparative experiments with Pisum, Vicia, and Helianthus primary root meristems demonstrated:

a) that cells may have 2 P.C.P. but the probability that they will arrest in either G1 or G2 is not always equal. For example, in Vicia there was a greater probability that a cell could stop in G2 rather than in G1,

b) that some species, Helianthus for example, may have meristematic cells with one predominent P.C.P.,

c) that some species, again exemplified by Helianthus, may have meristems with cell subpopulations that either do not have P.C.P. or are endowed with systems capable of satisfying the P.C.P. in spite of the effects of starvation and,

d) that even in cells of species with 2 P.C.P. the time needed to satisfy their requirements when provided sucrose was not the same. Both Pisum and Vicia had 2 P.C.P. but the former displayed a T_{ds} and T_{dm} of 11 hours while the latter had a T_{ds} of 7-8 and a T_{dm} of 7 hours. (Comparisons of this nature are only valid if meristems are deprived of carbohydrate for the same length of time, as T_{ds} and T_{dm} increase with the duration of starvation[21]).

In conclusion, some of the more general characteristics of the P.C.P. in Pisum are listed on the following page. Application of these characteristics to species other than Pisum, of course, should be done with caution and more appropriately accompanied by data. The list does however, provide a set of guidelines that are testable and useful for comparative studies.

SOME CHARACTERISTICS OF THE PRINCIPAL CONTROL POINTS IN *PISUM* ROOT MERISTEM CELLS

1. Located in G1 and G2 and function independently in the same cell.

2. "Sphere of influence" confined to G1 and G2.

3. Unidirectional, oriented toward mitosis.

4. Resolved when protein for the G1→S and G2→M transitions are deficient.

5. Require aerobic metabolism for expression.

6.. Sensitive to ionizing radiation.

7. Capable of recovery after damaged by ionizing radiation.

REFERENCES

1. Brown, R. (1951) The effects of temperature on the division of different stages of cell division in the root-tip. J. Exptl. Botany 2: 96-100.

2. Gray, L. H. and Scholes, M. E. (1951) The effect of ionizing radiations on the broad bean root. Brit. J. Radiol. 24: 82-92; 176-180; 228-236; 281-285; 348-352.

3. Gelfant, S. (1966) Pattern of cell division: the demonstration of discrete cell populations. In: Methods in Cell Physiology (D. M. Prescott, ed.), Academic Press Inc., New York.

4. Howard, A. and Pelc, S. R. (1953) Synthesis of deoxyribonucleic acid in normal and irradiated cells and its relation to chromosome breakage. Heredity 6 (Suppl.): 261-273.

5. Kovacs, C. J. and Van't Hof, J. (1969) Synchronization of a proliferative population in cultured root meristems and evidence for a G1 controlling site in the mitotic cycle. J. Cell Biol. 43 (Abstract):72a

6. Kovacs, C. J. and Van't Hof, J. (1970) Synchronization of a proliferative population in a cultured plant tissue: kinetic evidence for a G1/S population. J. Cell Biol. 47: 536-539.

7. Kovacs, C. J. and Van't Hof, J. (1971) Mitotic delay and the regulating events of plant cell proliferation: DNA replication by a G1/S population. Radiation Res. (In press).

8. Kovacs, C. J. and Van't Hof, J. (1971) Changes in polysome activity as a temporal expression of mitotic delay. Radiation Res. (Submitted).

9. Laughlin, H. H. (1917) The dynamics of cell-division. Soc. for Exptl. Biol. and Med. 15: 117-122.

10. Laughlin, H. H. (1919) Duration of the several mitotic stages in the dividing root tip cells of the common onion. Carn. Inst. Wash. Pub. 265.

11. Lea, D. E. (1955) Actions of Radiations on Living Cells. 2nd Edn. The University Press, Cambridge.

12. Quastler, H. (1963) The analysis of cell population kinetics. In: Cell Proliferation, (L. F. Lamerton and R. J. M. Fry, eds.), F. A. Davis Co., Philadelphia.

13. Quastler, H. and Sherman, F. G. (1959) Cell population in the intestinal epithelium of the mouse. Exptl. Cell Res. 17: 420-438.

14. Taylor, J. H., Woods, P. S. and Hughes, W. L. (1957) The organization and duplication of chromosomes as revealed by autoradiographic studies using tritium labeled thymidine. Proc. Natl. Acad. Sci. 43: 122-128.

15. Van't Hof, J. (1965) Cell population kinetics of excised roots of Pisum sativum. J. Cell Biol. 27: 179-189.

16. Van't Hof, J. (1966) Experimental control of DNA synthesizing and dividing cells in excised root tips of Pisum. Amer. J. Botany 53: 970-976.

17. Van't Hof, J. (1967) Recovery enhancement of G1 and G2 meristematic cells in excised pea roots during an induced post irradiation stationary phase. Radiation Res. 32: 792-803.

18. Van't Hof, J. (1968) Control of cell progression through the mitotic cycle by carbohydrate provision I. Regulation of cell division in excised plant tissue. J. Cell Biol. 37: 773-778.

19. Van't Hof, J. (1970) Metabolism and the prolonged retention of cells in the G1 and S period of the mitotic cycle of cultured pea roots. Exptl. Cell Res. 61: 173-182.

20. Van't Hof, J. (1970) Postirradiation Hypoxia and the recovery of G1 and G2 cells and DNA synthesis in cultured root meristems. Radiation Res. 41: 538-551.

21. Van't Hof, J. (1971) Two principal points of control in the mitotic cycle of pea meristem cells; energy considerations, characterization and radiosensitivity. In: Proc. Fourth International Congress of Radiation Research, (H. Duplan, ed.), Gordon and Breach Science Publishers, London. (In press).

22. Van't Hof, J. and Kovacs, C. J. (1970) Mitotic delay in two biochemically different G1 cell populations in cultured roots of pea (*Pisum sativum*). Radiation Res. 44: 700-712.

23. Webster, P. L. and Van't Hof, J. (1969) Dependence on energy and aerobic metabolism of DNA synthesis and mitosis by G1 and G2 cells. Exptl. Cell Res. 55: 88-94.

24. Webster, P. L. and Van't Hof, J. (1970) Initiation of DNA synthesis and mitotsis in stationary, transitional and proliferative phase meristems; requirements for RNA and protein synthesis. Amer. J. Botany 57: 130-139.

25. White, P. R. (1943) A Handbook of Plant Tissue Culture. Cattell, Lancaster, Pa.

26. Wimber, D. E. (1960) Duration of the nuclear cycle in *Tradescantia paludosa* root tips as measured with ^{3}H-thymidine. Amer. J. Botany 47: 828-834.

ACKNOWLEDGEMENTS

The authors wish to thank Dr. D. R. Burholt for helpful suggestions and discussions concerning the manuscript.

-DISCUSSION-

HABER - Did you check the effectiveness of the inhibition you used?

VAN'T HOF - Yes, our inhibitor experiments were accompanied by appropriate checks using labeled amino acids, RNA precursors, etc., and they were all found to be effective within the expected range for these substances.

D'AMATO - I am interested in the contrast you have found between *Vicia* and *Pisum* on one hand and *Helianthus* on the other; I think it reflects a much greater control of the synthesis of DNA and cell division in *Helianthus*. We find the same in our work with dry seed

D'AMATO - where both Vicia and Pisum have both 2C and 4C cells in the dormant meristem whereas Helianthus has only 2C.

VAN'T HOF - Yes; we can produce experimentally cell arrest that mimics that which occurs naturally and we have not found anything in our carbohydrate starvation studies that conflict with your findings in dry seed meristems.

D'AMATO - Further, similarly to your findings we can show that a G_1 population whilst uniform with respect to DNA and histone content per nucleus, analyzed in isolated nuclei, may not be uniform with regard to total protein. We think that one can have stoppage of DNA and histone synthesis whilst total protein continues to increase and this may be important in connection with the transition from chromosome to chromatid breakage response which occurs at the end of G_1.

VAN'T HOF - Even though some G_1 cells have more protein than others the proteins necessary for progression into S are lacking. Studies with labeled amino acids have shown that protein synthesis does occur even during carbohydrate starvation and that the rate is about half that of unstarved meristems. This implies that the additional rate we observe in unstarved roots is concerned with proteins vital for cell progression in the cycle.

LANGE - You quoted a 12-hour lifetime for the proteins. Does this fit in with any of the blockages you were getting?

VAN'T HOF - The estimate of 12 hours was obtained from proliferating cells exposing them to high concentrations of puromycin and determining the time taken for MI to reach zero. This agrees with the time taken for the MI to reach zero in excised roots. From our puromycin studies we have concluded that protein is always necessary for cell advancement regardless of where the cell is in the cycle. But the reason for the term "principle controlling point" is that though there are many factors that control cell division these must be subsidiary to the factors that block the cells in G_1 or G_2, otherwise stoppage would occur in other phases.

CONGER, A. - From your diagram I assume that cells stopped in G_2 by starvation are uniformly distributed throughout G_2, but what about the case of FUdR treatment?

VAN'T HOF - We believe that they are all stopped at exactly the same position, on the G_1-S border so that the frequency distribution on the diagram would be represented by a vertical line at that position. We have no direct evidence, however, that FUdR acts at the controlling point.

SMITH - What proportion of cells are killed in your roots by 300R?

VAN'T HOF - Fifteen percent of cells are killed by 300R. We cannot expose a meristem to any dose of ionizing radiation without killing cells.

CONGER, A. - At what time do they die?

VAN'T HOF - Death appears to occur at the stage of the cycle occupied by the cells at the time of exposure (i.e., it is interphase death).

TORREY - Could your analysis be carried out at a histological level?

VAN'T HOF - Yes; we are interested in this type of analysis to try to locate the "leakers" in *Helianthus*.

SAVAGE - Is the difference between *Helianthus* and the other species due to the fact that it may have an alternative energy supply, for example, from its abundant oil content?

VAN'T HOF - Yes. The "leakers" must either derive their energy from autolysis or from other cells. However, they and their progeny do not continue to progress in the cycle indefinitely.

MERISTEMATIC ACTIVITY IN RELATION TO WOUND XYLEM DIFFERENTIATION*

D. E. FOSKET

Department of Developmental and Cell Biology, University of California, Irvine, California, 92664, USA

INTRODUCTION

Many plant tissues and organs display a biphasic growth response. During the first phase, growth is brought about primarily by an increase in cell number, with little change in the average cell size. As the rate of cell division declines, the second mode of growth is exhibited in which the average cell size increases as many of the derivatives become characteristically differentiated, but these changes are accompanied by a comparatively small increase in cell number, if any. This developmental pattern is typical of root growth,[6,15,23] where the two growth processes tend to be spatially separated, and of leaf growth,[9,28,33] where the two processes tend to be separated in time.

The radial growth of the stem which is brought about by the vascular cambium also is biphasic. Cambial activity is similar to root growth in that the two growth phases usually occur simultaneously, but they are found in different tissue regions. The vascular cambium itself represents a cylindrical, meristematic tissue which is unique only in the elongate nature of most of the dividing cells. With the continued mitotic activity of the cambium, some of the derivatives become spatially removed from the actively dividing cells; they enter the second phase of their growth in which they stop dividing and enlarge radially while they become differentiated as vascular elements.

THE RELATIONSHIP BETWEEN CELL DIVISION AND CYTODIFFERENTIATION

What is the significance of this association between cell division and cytodifferentiation? Initially, cell division must function as the source of new cells which may undergo differentiation. To some extent this is obvious since a cell must come into existence before it can be differentiated. However, we want to know if this is the only significance of the cell division process to cytodifferentiation. Is a cell committed to a particular course of differentiation during the meristematic activity that produced it, or are cells only formed during meristematic activity with their fate to be determined later. Cells appear to be able to sense their position within a meristem, in relation to the tissues below them, and use this information to determine which of the various possible cell types they will become. The idea that positional information is an important factor in plant development is not new.[35] What we are asking at the present time is; when does a cell perceive its position? Does a cell discover where it is, and thus what its future function must be, during the mitotic event that produced it?

There isn't a simple answer to this question. In some cases the relationship between meristematic activity and cytodifferentiation is fortuitous. For example, many wound responses are biphasic in that a period of wound-induced meristematic activity precedes a number of cellular activities such as suberization, which may be considered to be examples of cytodifferentiation. Recently Kahl et al.[24] demonstrated that suberization would occur in potato tubers even though wound-induced mitotic activity was prevented. Apparently mature, non-dividing parenchyma cells can be induced to form suberin. Nevertheless, in many plant and animal systems cell division appears to be essential for the expression of differentiative potential in the sense that the initiation of a particular course of cytodifferentiation only occurs during some phase of the mitotic cycle.[1,2,14,17,18,32,34,36]

DNA SYNTHESIS OR CELL DIVISION?

It is important to distinguish between a differentiation process which is dependent upon DNA synthesis and one which is initiated during some phase of the mitotic cycle. These may be two entirely different phenomena. For example, specific inhibitors of DNA synthesis have been shown to prevent auxin-induced cell expansion in excised tobacco pith tissues[27] and the induction of flowering in Pharbitis.[37] Yet in both cases colchicine, which prevents cytokinesis without interfering with DNA replication, is not effective in blocking these processes. These results lead Zeevaart[37] to suggest that it was the DNA synthetic period of the mitotic cycle which was of significance for the induction of differentiation, while the remainder of the division cycle was incidental. This may well be the case. However, some recent findings have added further complexities to the relationship between DNA synthesis and cell differentiation, making it difficult to draw this distinction on the basis of inhibitor studies alone. A general inhibitor of DNA synthesis, such as 5-fluoro-deoxyuridine (FUdR), is as effective in blocking organelle DNA synthesis as nuclear DNA synthesis. Some types of cytodifferentiation, cell elongation for example, may be dependent upon organelle DNA synthesis. Nitsan and Lang[29,30] and Bopp[34] have used FUdR, along with other inhibitors of DNA synthesis, to demonstrate that DNA synthesis can be a requirement for cell elongation. Recently Degani and coworkers[12,13] found that it was not nuclear DNA synthesis, but mitochondrial DNA synthesis which accompanied and was necessary for gibberellin-induced elongation in cucumber seedlings.

In addition to organelle DNA synthesis, it is possible that the DNA synthesis accompanying a given developmental process may represent the amplification of genes specific for that process. Recent work on the amplification of the ribosomal cistron during amphibian egg development[5] has established that a fraction of the total cell genome may be synthesized, utilized and destroyed to accomplish a specific cellular goal, independent of the normal cycle of chromosome replication and division. Although specific gene amplification has not been shown to occur in plants, Chen and Osborn[7] have described a modification of the physical properties of wheat DNA during the germination of the seed which could be explained if one assumed that some of the DNA coding for ribosomal RNA was lost during germination. Obviously, if gene amplification should prove to be a concomitant of a particular plant developmental process, it also may be found that this process is dependent upon DNA synthesis, independent of cell division or the DNA synthesis associated with the mitotic cycle.

In plants and insects endomitosis and polytene may further complicate the relationship between cytodifferentiation and DNA synthesis. Endopolyploidy has been shown to accompany tracheary element formation in many species[26] and trichoblast formation in *Hydrocharis* roots,[8] as well as many other plant developmental processes.[10,11] Kafatos and Feder[25] have described a remarkable case in silkmoths where generalized epidermal cells in the galea undergo successive endomitotic reduplications as they differentiate into secretory cells and produce large quantities of the enzyme cocoonase. The significance of polyploidy to cytodifferentiation is not known. As Cutter and Feldman[8] have pointed out, it is unlikely that polyploidy is required for differentiation, at least in the case of root hair formation, since the trichoblasts of many species appear to be diploid. The fact that the trichoblasts of

Hydrocharis are unusually large suggests that the endomitotic reduplications accompanying their formation could represent another form of gene amplification.

CELL DIVISION AND WOUND XYLEM DIFFERENTIATION

Despite these difficulties it has been established that some types of plant cytodifferentiation are initiated during the mitotic cycle. Although this phenomenon has been investigated in a variety of plant tissues, the nature of the evidence indicating this relationship exists can be illustrated by considering the formation of tracheary elements in wounded Coleus stems and cultured internodal segments.

1. Tracheary Element Formation in Cultured Coleus Internodes.

When 2 mm thick segments of the 2nd internode of Coleus are placed in axenic culture on a simple medium containing only 2% sucrose (SA) or 2% sucrose and 0.05 ug/l indoleacetic acid (SAIAA), the segments will be found to contain a number of newly differentiated tracheary elements after 7 days in culture. These tracheary elements have the morphological characteristics of wound vessel members.[19] An examination of the time course of wound vessel member formation in the cultured stem segments has revealed that there is a considerable time lag before the first differentiating tracheary elements can be observed (Fig. 1). Although the incorporation of auxin into the culture medium greatly increases the number of wound vessel members formed in 7 days, it does not shorten the lag period; the first wound vessel members appear after 3 days in culture, whether the stem segments are given auxin or not. Most of the tracheary elements formed by the cultured stem segments appear between the 3rd and the 5th day of culture. Comparatively few tracheary elements are formed after the 5th day.

Although the 1st 3 days of the culture period represent a lag period for xylem formation, it has proved to be the most active period for DNA synthesis, as revealed by the incorporation of ^{3}H-thymidine into DNA (Fig. 1). The time course of ^{3}H-thymidine incorporation into DNA is a reflection of the mitotic activity of the cultured stem segments (Table I). In either case, the respective process reached a maximum at approximately 48 hours after segment isolation from a comparatively low value observed on the 1st day of culture. Both mitotic activity and DNA synthetic activity declined from the peak observed at day 2 until, by the 5th day of culture, the values observed were similar to those found on the 1st day. Thus, the cultured stem segments exhibit a biphasic growth response in which a period of mitotic activity tends to precede cytodifferentiation.

FIGURE LEGENDS

Figure 1. The time course of DNA synthesis and xylem differentiation in Coleus stem segments cultured in vitro. Counts of differentiating wound xylem elements were made at daily intervals in stem segments cultured on SA or SAIAA medium. The mean number of wound vessel members per segment for each daily sample is given. Daily rates of DNA synthesis were estimated from the amount of ^{3}H-thymidine incorporated into DNA within a 12 hour labeling period. The observed incorporation of ^{3}H-thymidine in SAIAA-grown stem segments is indicated by the vertical bars, the width of which represents the duration of the labeling period.

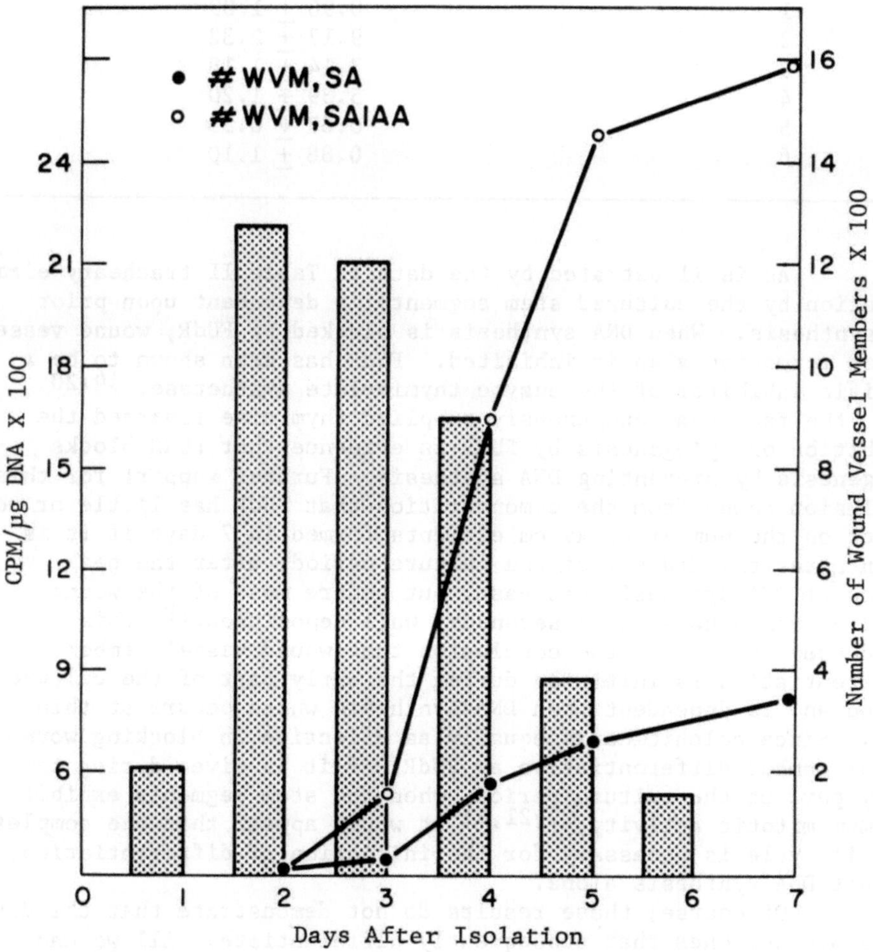

Table I. MITOTIC ACTIVITY IN CULTURED COLEUS STEM SEGMENTS. Using aseptic techniques, tissue slices from the 2nd internode were isolated on sucrose-agar medium containing 0.05 ug/l IAA. At daily intervals segments were fixed in 3:1 fixative, Feulgen stained, treated with pectinase, and squashed onto glass slides. The number of metaphase figures per 1000 cells was determined for each segment. The average number of metaphase figures per thousand cells is given, together with the standard deviation of the mean.[21]

Time after isolation (days)	Number of metaphase figures per 1000 cells
1	0.96 ± 1.09
2	9.17 ± 2.32
3	7.14 ± 1.18
4	5.39 ± 1.20
5	0.87 ± 0.53
6	0.88 ± 1.10

As is illustrated by the data in Table II tracheary element formation by the cultured stem segments is dependent upon prior DNA synthesis. When DNA synthesis is blocked by FUdR, wound vessel member formation also is inhibited. FUdR has been shown to be a specific inhibitor of the enzyme thymidylate synthetase.[16,20] Thus, the fact that exogenously supplied thymidine reversed the inhibition of xylogenesis by FUdR is evidence that FUdR blocks xylogenesis by preventing DNA synthesis. Further support for this conclusion comes from the demonstration that FUdR has little or no effect on the number of xylem elements formed in 7 days if it is given after the 3rd day of the culture period, after the peak period of DNA synthesis has past, but before most of the wound xylem elements have begun secondary wall deposition.[18] This would seem to justify the conclusion that wound vessel member differentiation is initiated during the early part of the culture period and is dependent upon DNA synthesis which occurs at this time. Since colchicine is equally as effective in blocking wound vessel member differentiation as FUdR, if it is given during the early part of the culture period, when the stem segments exhibit maximum mitotic activity,[17,21,31] it would appear that the complete mitotic cycle is necessary for the initiation of differentiation, and not DNA synthesis alone.

Of course, these results do not demonstrate that the dividing cells are the ones that subsequently differentiate. All we can conclude from the inhibitor studies is that DNA synthesis and cell division must occur in the cultured stem segments for tracheary elements to be formed. It is entirely possible that dividing cells

Table II. THE EFFECT OF FUdR ON WOUND XYLEM DIFFERENTIATION IN CULTURED COLEUS STEM SEGMENTS. Segments were isolated on SA medium, SAIAA medium, or these same media supplemented with 10^{-6} M FUdR and 10^{-5} M Uridine. After 7 days in culture segments were cleared and squashed for counts of differentiated wound vessel members. The means are given together with the standard deviation of the means.[18]

Medium	No. of WVMs	% Inhibition
SA	514 ± 44	-
SA+10^{-6}M FUdR	106 ± 23	80
SA+10^{-6}M FUdR+10^{-5}M Thymidine	489 ± 31	5
SAIAA	1877 ± 147	-
SAIAA+10^{-6}M FUdR	382 ± 32	79
SAIAA+10^{-6}M FUdR+10^{-5} Thymidine	1746 ± 97	7

produce substances which are necessary for the differentiation of contiguous, non-dividing cells. This possibility has been examined, using histological sections and autoradiography, to determine where mitotic activity occurs and what relation this activity bears to tracheary element differentiation in the cultured *Coleus* stem segments.

The 2nd internodes of *Coleus* used in these studies, defined according to Jacobs'[22] criteria, average 1.25 ± 0.18 cm in length. In cross section, the xylem tissue of each of the four main vascular bundles contains an average of 410 ± 34 cells, of which 29 ± 1.26 are annular or spirally-thickened vessel elements. The remaining cells are thin-walled xylem parenchyma. The xylem tissue of each of the major bundles tends to be arranged in radial rows, with the most recently formed vessels nearest the phloem. In contrast to the xylem, the phloem is not arranged in radial rows, but consists of an average of 11 sieve tube-containing clusters (Fig. 2). An additional 5-6 sieve tube-containing clusters are found in the interfascicular region, with vessels associated with one to three of these phloem bundles. The small vascular bundle midway between two large corner vascular bundles was found to contain 3-5 vessels, while the procambial tissue between the vessels and the phloem of such small bundles was in radial rows, 2-3 cells wide. The 2nd internode does not have an interfascicular cambium (Fig. 9).

Segments of the 2nd internode, when cultured for 7 days on SA or SAIAA media, exhibit a characteristic swelling at their basal ends. This swelling was associated with the morphological basal end regardless of the orientation of the segment with respect to the culture medium. An analysis of histological sections and autoradiographs has demonstrated that this basal swelling is accompanied by the proliferation of the xylem parenchyma, procambial

FIGURE LEGENDS

Figures 2-10--Fig. 2.-A normal 2nd internode in X.S. showing one of the 4 corner vascular bundles. Fig. 3.-Autoradiograph of a cultured 2nd internodal segment which had received 2uc/ml ^{3}H-thymidine at 60-72 hrs after culturing. Fig. 4.-An autoradiograph of an X.S. taken 560 u from the base of the cultured stem segment shown in the previous figure. Fig. 5.-An autoradiograph of an X.S. taken at 1440 u from the basal end of the cultured stem segment shown in figure 3. Fig. 6.-An X.S. taken at 580 u from the basal end of a segment cultured for 7 days on SAIAA medium. Fig. 7.-An X.S. taken at 1220 u from the basal end of a cultured stem segment after 7 days. Fig. 8.-A cross section taken from near the apical end of a segment cultured for 7 days on SAIAA medium. Fig. 9.-An X.S. of a 2nd internode before culturing showing the anatomy of the interfascicular region. Fig. 10.-An autoradiograph of an X.S. taken 440 u from the base of a cultured segment, after receiving ^{3}H-thymidine for 12 hours during the period, 60-72 hrs after culturing.

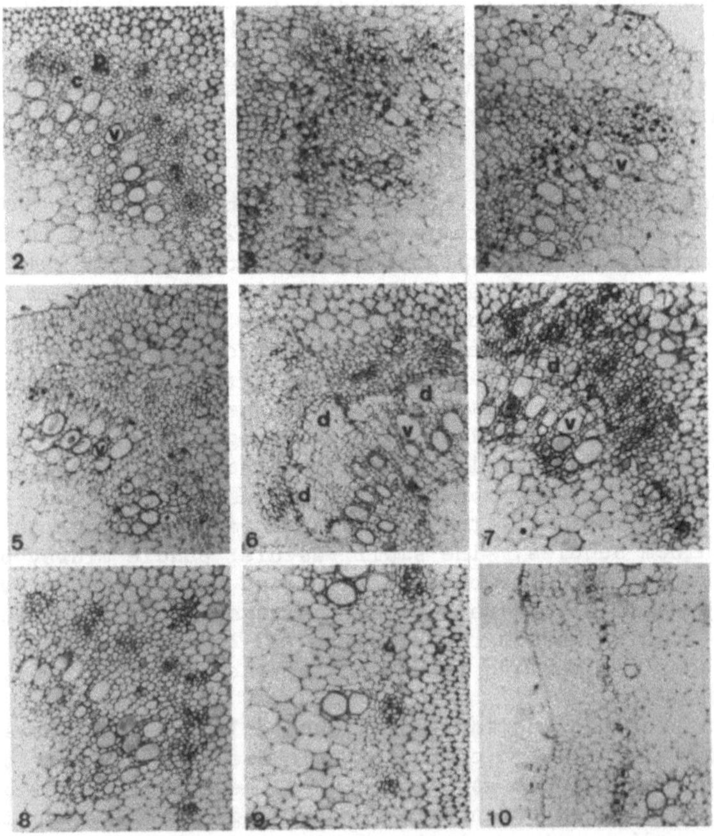

cells and phloem, as well as the sporatic activation of the vascular cambium. Figures 3 through 5 are autoradiographs of sections cut from stem segments which had been given ^{3}H-thymidine for a 12 hour period, 60-72 hours after the start of culturing. The section shown in Fig. 3 was 300 u from the base of the segment while Fig. 4 was a section 560 u from the base and Fig. 5 was 1440 u from the base of a segment whose total length was 3270 u. Labeled nuclei largely are confined to the cells of the vascular bundles in all three sections, but progressively fewer cells have labeled nuclei with increasing distance from the segment base. This labeling pattern tends to be reflected in the pattern of wound vessel member differentiation, as seen in cross sections taken from segments cultured for 7 days. The most extensive wound vessel member formation is in association with the main vascular bundles near the base of the stem segment where the differentiated cells are formed from derivatives of the procambial cells (Fig. 6). Fewer tracheary elements are produced in the middle of the stem segments (Fig. 7), but even here the newly formed xylem elements tend to be differentiated from procambial cells which have divided during the culture period. Wound vessel members are not commonly found in the apical third of the cultured segments (Fig. 8).

The cultured stem segments also display a limited and sporadic development of an interfascicular cambium. As is shown by the autoradiograph in Fig. 10, not every cell is labeled in the boundary layer between the pith and the cortex which gives rise to the interfascicular cambium. Instead, activity tends to be confined to the cells in this region near the existing vascular bundles. After 7 days in culture, these same regions show evidence of fairly extensive cambial activity, with many of the cambial derivatives having differentiated as wound vessel members (Fig. 11). As with the wound vessel member formation associated with the 4 main vascular bundles, the xylem elements formed from the interfascicular cambium, and the development of the cambium itself, tend to be confined to the basal half of the stem segments (Fig. 12).

Although most of the cells incorporating ^{3}H-thymidine into DNA were associated with the vascular tissues or the developing interfascicular cambium, a few labeled cells were always observed among the epidermal, cortical and pith parenchyma cells. Most of these cells did not subsequently differentiate as wound vessel members. No wound vessel members have been observed in either the epidermis or cortex. Occasionally a nest of wound vessel members is found in the pith region, near the basal ends of one or more of the 4 main vascular bundles (Fig. 13). There is little doubt that many, if not all, of these wound vessel members represent derivatives of pith parenchyma cells which divided during the culture period.

FIGURE LEGENDS

Figures 11-19--Fig. 11-An X.S. of a 2nd internodal segment cultured 7 days on SAIAA medium. Fig. 12-An L.S. through the interfascicular region of a 2nd internodal segment cultured 7 days on SAIAA medium. The basal end is facing up. Fig. 13-An X.S. taken at 230 u from the basal end of a segment cultured 7 days on SAIAA medium. Fig. 14-A cleared whole mount of a 2nd internode prepared 7 days after it had received a wound which severed one of the minor vascular bundles. Fig. 15-A cleared whole mount of a 5th internode as seen 7 days after receiving a single lateral wound. Fig. 16-A wounded 8th internode as seen in face view in a cleared whole mount prepared 7 days after wounding. Fig. 17-A fresh free hand section from a 2nd internode 7 days after wounding. Fig. 18-A fresh free hand X.S. from a 2nd internode, 7 days after wounding. Fig. 19-A fresh free hand X.S. of a 2nd internode taken below the wound. 7 days after wounding.

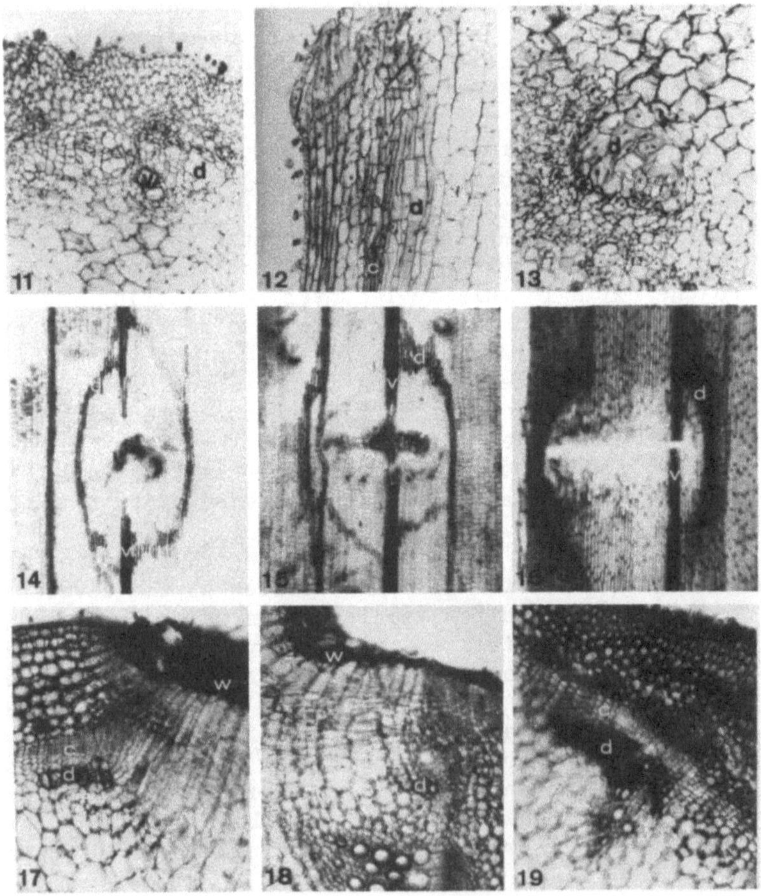

2. Cambial Activity and Wound Xylem Differentiation.

These results clearly demonstrate that most of the tracheary elements formed by the Coleus stem segments are cells which came into existence through mitotic activity during the culture period. Furthermore, most of the differentiated cells are derivatives of the procambium (or fascicular cambium) and the interfascicular cambium. This was somewhat unexpected since the interfascicular cambium normally is formed in the 4th or 5th internode of Coleus. It takes approximately 3 weeks for a 2nd internode to become the 5th internode on a plant growing under greenhouse conditions. This raises several questions: Is the developmental pattern we have observed in cultured 2nd internodal segments a modification of the normal stem wound response? Does wounding induce or stimulate cambial activity? Do the wound vessel elements formed around a stem wound in Coleus represent cambial derivatives? The answer to all these questions seems to be yes.

Figures 14, 15 and 16 are photomicrographs of cleared whole mounts of wounded 2nd, 5th and 8th internodes, respectively, as seen in face view. In each case a single lateral incision was made which severed one or more of the small vascular bundles between two large corner bundles, without rupturing the corner bundles. The photographs depict the pattern of xylem differentiation around the wound, 7 days after wounding. In all 3 cases tracheary elements have differentiated in an arc around the wound, joining the severed vascular bundle(s) both above and below the wound. Figures 17, 18, 20 and 23 are cross sections of wounded 2nd, 5th and 8th internodes, taken through the wound region and showing the developmental pattern 7 days after wounding. In all three cases wounding has induced mitotic activity in the pith adjacent to the wound where the plane of division tends to be oriented parallel to the wound cicatrix. Few if any tracheary elements were formed from these pith parenchyma derivatives. Instead, wound xylem elements are formed from cambial derivatives or cells immediately adjacent to cambial derivatives. An interfascicular cambium was formed in the 2nd internodes in the immediate vicinity of the wound. The degree of cambial activity exhibited by this newly formed interfascicular cambium varies considerably, even within the wound region on the same internode. Figure 17 illustrates a wounded 2nd internode with a well developed interfascicular cambium and numerous wound vessel members which have differentiated from cambial derivatives. Figure 18 illustrates another 2nd internode, but with a poorly developed interfascicular cambium and few wound vessel members at approximately the same position with respect to the wound as the 2nd internode shown in Figure 17. Almost all the wounded 2nd internodes displayed vigorous cambial activity in the interfascicular region immediately adjacent to the minor bundle which was severed by the wound, at the elevation where the newly formed wound tracheary elements rejoin it. These wound tracheary elements were derivatives of the interfascicular cambium which had developed in this region (Figure 19). The interfascicular cambial development in association with a minor vascular bundle below the wound was quite similar to the response observed in the basal halves of the cultured internodal segments (compare Figure 11 and Figure 19). Also, it is similar in kind to the responses of older Coleus internodes which have an interfascicular cambium at the time of wounding.

The section shown in Figure 21 was taken from below the wound in a 5th internode, at the point where the newly formed tracheary elements join the vascular bundle which was ruptured by the wound. Wound vessel members, which can be found among the rows of cambial derivatives, represent the only tracheary elements which have matured from these derivatives. The cambium and its derivatives in the vicinity of the wound averages about twice as many cells per file as the interfascicular cambium in the same internode, but on the side of the stem opposite the wound (Figure 22). Thus, wounding has accelerated cambial activity in the vicinity of the wound, much as it induced the formation of the interfascicular cambium in the 2nd internode. Cambial activity also is accelerated in the vicinity of a stem wound in the 8th internode, although the development of the wound response is somewhat more complex than was found to be the case in younger internodes. Figure 24 illustrates cambial activity and its relation to wound xylem formation in an 8th internode. The section was taken below the wound, but above the point at which the newly formed wound xylem elements rejoin the severed minor bundle. Figure 25 illustrates the development of the cambium and its derivatives in the same internode, but on the side of the stem opposite the wound. Near the wound the cambium and its derivatives contain an average of approximately 50% more cells per file than can be found in these same tissues on the opposite side of the stem. Wounding an 8th internode not only leads to a local stimulation of cambial activity, but results in the production of a more complex xylem tissue from the interfascicular cambium so stimulated. Normally the interfascicular cambium in Coleus only produces xylem parenchymal cells, which develop a uniformly-thickened, lignified secondary wall, but retain their protoplast at maturity and appear to function in starch storage. Only within the sectors defined by the original 4 corner bundles, plus a variable number of minor bundles, will vessel elements differentiate from some of the cambial derivatives. However, in the vicinity of a stem wound, the derivatives of the interfascicular cambium differentiate as a more heterogeneous tissue, consisting of wound vessel members as well as xylem parenchyma.

In summary, these results not only demonstrate that cell division is necessary for wound tracheary element formation, but that wound xylem elements are derived from cells that have divided in response to experimental treatment. Furthermore, the wound

FIGURE LEGENDS

Figures 20-25--Fresh free hand cross sections of wounded <u>Coleus</u> internodes taken 7 days after wounding. Fig. 20-Section taken from a wounded 5th internode through the wound region. Fig. 21-A section taken below the wound in a 5th internode. Fig. 22-Section taken from a wounded 5th internode on the side of the stem opposite the wound. Fig. 23-Section taken through the wound in an 8th internode. Fig. 24-An X.S. of an 8th internode taken below the wound. Fig. 25-The interfascicular cambium and its derivatives on the side of the stem opposite the wound in an 8th internode.

Explanation of symbols: d = newly differentiated xylem elements; v = pre-existing vessel or vessel element; c = cambium or pre-cambium; w = wound cicatrix.

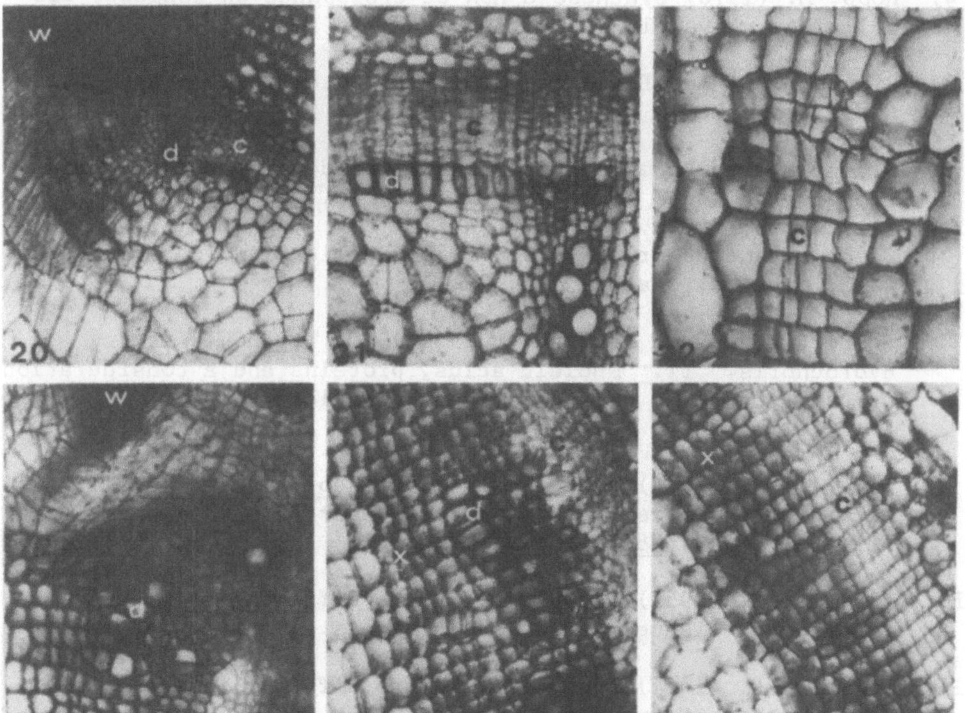

vessel members which are formed in cultured Coleus stem segments or in response to stem wounds, are derivatives of the vascular cambium. In some cases, particularly in response to wounding, cells in tissues other than the vascular cambium are induced to divide, but relatively few of their derivatives differentiate as tracheary elements.

REFERENCES

1. Bischoff, R. and Holtzer, H. (1969) Mitosis and the processes of differentiation of myogenic cells in vitro. J. Cell Biol. 41:188.

2. Bischoff, R. and Holtzer (1970) Inhibition of myoblast fusion after one round of DNA synthesis in 5-bromodeoxyuridine. J. Cell Biol. 44:134.

3. Bopp, M. (1968) Hemmung des Streckungswachstums etiolierter Sprossachen durch FUDR. Z. Pflanzenphysiol. 57:173.

4. Bopp, M. (1970) Synthese d'ADN et croissance en longueur des cellules. Physiol. Veg. 8:215.

5. Brown, D. D. and David, I. B. (1968) Specific gene amplification in oocytes. Science 160:272.

6. Brown, R. and Broadbent, D. (1950) The development of cells in the growing zones of the root. J. Exptl. Bot. 1:249.

7. Chen, D. and Osborne, D. J. (1970) Replication in germinating wheat embryos. Nature 225:336.

8. Cutter, E. G. and Feldman, L. J. (1970) Trichoblasts in Hydrochoris. II. Nucleic acids, proteins and a consideration of cell growth in relation to endopolyploidy. Amer. J. Bot. 57:202.

9. Dale, J. E. (1964) Leaf growth in Phaseolus vulgaris. I. Growth of the first pair of leaves under constant conditions. Ann. Bot. N. S. 28:579.

10. D'Amato, F. (1952) Polyploidy in the differentiation and function of tissues and cells in plants. Caryologia 4:311.

11. D'Amato, F. (1964) Endopolyploidy as a factor in plant tissue development. Caryologia 17:41.

12. Degani, Y. and Atsmon, D. (1970) Enhancement of non-nuclear DNA synthesis associated with hormone-induced elongation in the cucumber hypocotyl. Exptl. Cell Res. 61:226.

13. Degani, Y., Atsmon, D. and Halevy, A. (1970) DNA synthesis and hormone induced elongation in cucumber hypocotyl. Nature 228:554.

14. Eisenberg, S. and Yamada, T. (1966) A study of DNA synthesis during the transformation of the iris into lens in the lentectomized newt. J. Exptl. Zool. 162:353.

15. Erickson, R. O. and Goddard, D. R. (1951) An analysis of root growth in cellular and biochemical terms. Growth 10:89.

16. Flamm, W. G. and Birnstiel, M. L. (1964) Inhibition of DNA replication and its effect on histone synthesis. Exptl. Cell Res. 33:616.

17. Fosket, D. E. (1968) Cell division and the differentiation of wound-vessel members in cultured stem segments of Coleus. Proc. Natl. Acad. Sci. (U.S.) 59:1089.

18. Fosket, D. E. (1970) The time course of xylem differentiation and its relation to deoxyribonucleic acid synthesis in cultured Coleus stem segments. Plant Physiol. 46:64.

19. Fosket, D. E. and Roberts, L. W. (1964) Induction of wound-vessel differentiation in isolated Coleus stem segments in vitro. Amer. J. Bot. 51:19.

20. Hartman, K. V. and Heidelberger, C. (1961) Studies on fluorinated pyrimidines XII. Inhibition of thymidylate synthetase. J. Biol. Chem. 236:3006.

21. Hepler, P. K. and Fosket, D. E. (1971) The role of microtubules in vessel member differentiation in Coleus. Protoplasma (In press).

22. Jacobs, W. P. (1952) The role of auxin in differentiation of xylem around a wound. Amer. J. Bot. 39:301.

23. Jensen, W. A. (1955) A morphological and biochemical analysis of the early phases of cellular growth in the root tip of Vicia faba. Exptl. Cell Res. 8:506.

24. Kahl, G., Rosenstock, G. and Lang, H. (1969) Die Trennung von Zellteilung und Suberinsynthese in deteprimiertem pflanzlichem speichergewebe durch Tris-(hydroxymethyl) aminomethane. Planta 87:365.

25. Kafatos, F. C. and Feder, N. (1968) Cytodifferentiation during insect metamorphoses: The galea of silkmoths. Science 161: 470.

26. List, A., Jr. (1963) Some observations on DNA content and cell nuclear volume growth in the developing xylem cells of certain higher plants. Amer. J. Bot. 50:320.

27. Makeshwari, M. G. and Nooden, L. D. (1971) A requirement for DNA synthesis during auxin induction of cell enlargement in tobacco pith tissue. Physiol. Plant. 24:282.

28. Maksymowych, R. and Kettrick, M. A. (1970) DNA synthesis, cell division, and cell differentiation during leaf development of Xanthium pennsylvanicum. Amer. J. Bot. 57:844.

29. Nitsan, J. and Lang, A. (1965) Inhibition of cell division and cell elongation in higher plants by inhibition of DNA synthesis. Develop. Biol. 12:358.

30. Nitsan, J. and Lang, A. (1966) DNA synthesis in the elongating, mondividing cells of the lentil epicotyl and its promotion by gibberellin. Plant Physiol. 41:965.

31. Roberts, L. W. and Baba, S. (1968) IAA-induced xylem differentiation in the presence of colchicine. Plant and Cell Physiol. 9:315.

32. Stockdale, F. E. and Topper, Y. J. (1966) The role of DNA synthesis and mitosis in hormone-dependent differentiation Proc. Natl. Acad. Sci. (U.S.) 56:1283.

33. Sunderland, N. (1960) Cell division and expansion in the growth of the leaf. J. Exptl. Bot. 11:68.

34. Torrey, J. G. and Fosket, D. E. (1970) Cell division in relation to cytodifferentiation in cultured pea root segments. Amer. J. Bot. 57:1072.

35. Vöchting, H. (1878) Uber Organbildung im Pflanzenreich. Max Cohen, Bonn.

36. Voytovich, A. E. and Topper, Y. J. (1967) Hormone-dependent differentiation of immature mouse mammary gland in vitro. Science 158:1326.

37. Zeevaart, J. A. D. (1962) DNA multiplication as a requirement for expression of floral stimulus in Pharbitis nil. Plant Physiol. 37:296.

* The author would like to thank Dr. John G. Torrey, Harvard University, in whose laboratory this study was conducted.

-DISCUSSION-

CONGER, A. - Why do these Coleus segments only make vessels at the lower end?

FOSKET - It must be remembered that we are only looking at the remnant of a complete system in these segments. No matter how exogenous auxins are applied they all end up at the bottom end due to polar transport; in the cut segment there is nowhere for them to go and so they build up at the lower end.

HABER - Most of the vessels were formed in these systems following cell divisions; have you any comment on the minority that form vessels without division?

FOSKET - It may be that what we are observing is the completion of the differentiation process which was begun before the stem segment was isolated. Usually the few vessel elements formed in the absence of cell division are right in vascular bundle.

BALL - You mention auxins and other substances from which I infer that you do not think that auxin is the main substance causing the differentiation of secondary wall.

FOSKET - In Coleus it is clear that auxin is the principle factor limiting differentiation. However, in other systems it is clear that other substances are also involved; in pea root segments, for example, Torrey has shown that there is an absolute requirement for cytokinin. The limiting factor differs in different systems. This isn't to say that auxin isn't necessary for differentiation in pea root segments and cytokinin in Coleus stem segments. A variety of substances and conditions are necessary for differentiation, any one of which could become limiting.

In Coleus we can prevent differentiation entirely by omitting sucrose from the culture medium.

BALL - Is the response to wounding the production of additional auxin?

FOSKET - I don't know, but I would assume that cytokinins are produced.

BALL - In the normal system of differentiation the nucleus generally becomes highly polyploid before it disappears; does this happen in your segments and what would be the effect of colchicine?

FOSKET - Wound vessel members do become polyploid. Colchicine does not block that response, but FUdR would. FUdR inhibits differentiation but this result by itself does mean that cell division alone is

FOSKET - necessary for xylem formation. Other interpretations are possible; for example, in tobacco auxin-induced cell expansion is DNA synthesis dependent but not cell division dependent. Also, in the flowering response, it has been shown that DNA synthesis is necessary for the induction of flowering but that cell division is not. So these systems are quite different from xylogenesis in the Coleus segments where cell division is also necessary. Whether polyploidy also is required for tracheary element formation we can't say.

BALL - Polyploidization must be very rapid if it precedes secondary wall formation.

FOSKET - Secondary thickening occurs very rapidly in wound vessel elements but very slowly in xylem parenchyma. Since the xylem parenchyma nuclei remain diploid there may be some connection between rate of wall deposition and degree of polyploidy of the nucleus.

D'AMATO - In lettuce cotyledons Mrs. V. Nuti investigating the effects of auxins and kinetin on adventitious buds has shown the formation of some vascular elements without any previous mitosis (Giorn. Bot. Ital., 104:443-455, 1971). Extensive vascular differentiation is necessary in the cotyledon before the epidermal division which produces the adventitious buds can occur. Mrs. Nuti thinks that the sequence of events is the amplification of cistrons coding for ribosomal RNA, followed by DNA doubling and then mitosis. Ribosomal RNA may be an important prerequisite for differentiation. Miss S. Avanzi investigating metaxylem differentiation in Allium root, finds the following sequence of events: (1) endopolyploidy, (2) extra synthesis of DNA only in nucleolar regions forming a large mass of ribosomal DNA, (3) extrusion of this material into the cytoplasm (Caryologia, in press). I am inclined to think that cell division is not an absolute prerequisite.

FOSKET - I did not want to give the impression that tracheary elements cannot be produced from pith cells but only that the pith is not the usual source of wound vessel members. Comer has found that large numbers of wound vessel members can be induced to form in isolated Coleus pith explants under the proper cultural conditions (personal communication). He has found that cell division precedes xylogenesis in these pith explants, although we don't know if cell division is necessary for differentiation in this system at the present time. The stem explant work suggests that cell division would be necessary for the initiation of xylogenesis from Coleus pith cells, however. It is also quite likely in pith that a good deal of macromolecular synthesis would have to precede cell division because these are quiescent cells in Coleus and normally would never divide.

A MORPHOGENETICIST'S VIEW OF CORRELATIVE INHIBITION IN THE SHOOT*

ELIZABETH G. CUTTER

Department of Botany, University of California

Davis, California 95616, U.S.A.

INTRODUCTION

The complexity of shoot form depends, in part, on a number of correlative effects between various organs. The best known of these is apical dominance, the effect of the terminal apex on the lateral buds, but there are also correlative effects between the lateral buds themselves, and growing leaves or leaf primordia may affect other leaves or the axillary buds which they subtend. Correlative inhibition between buds has been recognized most frequently in the case of cotyledonary buds (Snow[44]; Sachs[36]; McIntyre[24]), but exists also between two buds at higher nodes in decussate or bijugate species, at least if the buds are of unequal size. In most species with paired leaves the buds may possess different developmental potentialities, one of each pair developing as a vegetative shoot and the other as a flower, inflorescence or thorn; in others they merely develop as shoots of unequal size (Champagnat[5]; Goebel[13]; Loiseau[21]; Raciborski[35]). Such species provide good material for the study of the less common correlative effects in the shoot. Various experiments have been or will be reported in more detail elsewhere (Cutter[8]; Cutter and Chiu[9]), but are discussed briefly now along with a more general consideration of the morphogenetic aspects of the phenomenon of apical dominance.

Although apical dominance, the inhibition of lateral shoots by the main apex, is one of the most evident and widespread of growth phenomena in plants, it is still very incompletely understood. The relevant literature, which has been recently reviewed by Phillips[32], is extensive and quite confusing. Various current theories attribute the mechanism of apical dominance to the direct or indirect action of auxin synthesized by the main shoot apex, competition between the main apex and the lateral buds for nutrients, competition for a limited water supply, inadequate or incomplete differentiation of the vascular trace between the bud and the main stem, and the balance between auxin and cytokinin at the bud locus.

In different species, or even occasionally in the same species under different conditions, apical dominance may be complete or only partial; in shoots with partial apical dominance lateral branches grow out but their growth is still inhibited by the main shoot apex, whereas if apical dominance is complete no outgrowth of the lateral buds occurs. In 1937 Snow[45] considered whether the correlative inhibition of growing shoots might be brought about by different means from that of buds, and in particular whether the "direct" theory might apply only to the latter. He concluded that probably no such distinction existed.

Nevertheless, structures influenced by apical dominance may range from bud meristems with no leaf primordia in the ferns Onoclea and Matteuccia (Wardlaw[49,50]) to growing leafy shoots in Coleus (Jacobs, Danielson, Hurst and Adams[17]), and it is reasonable to suppose that rather different facets of shoot morphogenesis, and different controlling factors, may assume greater relative importance in these two examples. Although cell division in the apical meristem, leaf formation, leaf expansion and internodal extension, aspects of shoot morphogenesis which are controlled by a number of different factors, may all ultimately be involved in both instances, in Coleus these must be already occurring in slowly growing "inhibited" shoots, and presumably any or all of them could be stimulated by release from apical dominance, whereas in released buds of Onoclea and Matteuccia they must occur in proper sequence or not at all. A factor or factors promoting cell division in the apical meristem would thus assume fundamental importance in the control of bud outgrowth in the latter species. Thus, as indeed the composite literature suggests, even within a species the mechanism of apical dominance may well be attributable not to one, but to several different factors, any one of which may assume greater relative importance at different stages of morphogenesis of the lateral shoot. The possible factors involved are considered below, in relation to current experimental results and what is known of lateral bud morphogenesis.

MATERIALS AND METHODS

A species of Hygrophila (Acanthaceae) with complete apical dominance was studied. At each node of the decussate shoot a larger (+) and a smaller (-) bud were subtended; the large buds were arranged in a helix up the stem, the smaller buds forming a second such helix (Fig. 1). Nodes of Hygrophila were surface-

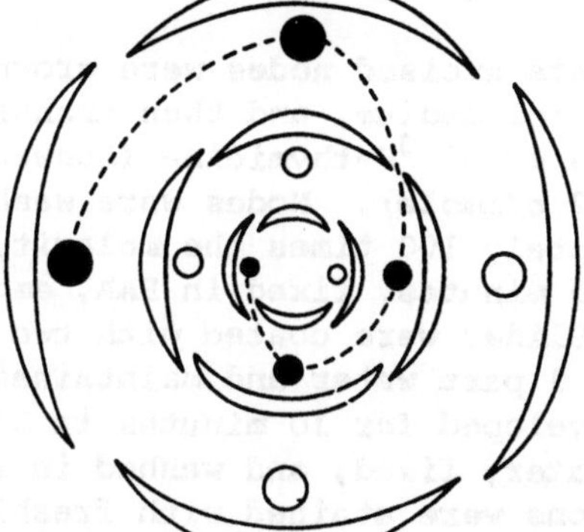

Figure 1. Diagram showing the arrangement of larger or + buds (solid circles) and smaller or - buds (open circles) in a helicoidal system. (From Cutter[6], Fig. 14.2B, p. 225; after Goebel[13]).

sterilized and placed in sterile aqueous nutrient medium, and a clone was established by subculturing individual nodes from a lateral shoot which remained sterile. The control medium was 1/2-strength Knop's solution (Gautheret[12]), with Nitsch's[28] microelements and 2% sucrose, autoclaved at 15 lb pressure for 20 minutes. Cultures were grown in test tubes in a culture room at 21°C with 14 hr light of 360 foot-candles, or occasionally in a growth chamber at 23°C with 16 hr light of approximately 2000 foot-candles.

In some experiments single nodes were split longitudinally under aseptic conditions, and selected half-nodes were grafted together by tying in several places with pieces of nylon fishing line. This technique is described in more detail elsewhere (Cutter and Chiu[9]). Sucrose was sometimes omitted from the medium for grafted nodes, in case sterility had not been adequately maintained.

Various hormonal treatments were applied. Hormones were usually added to the nutrient medium at the concentrations stated prior to autoclaving. In some experiments a solution of the hormone in 1% or 0.8% agar was placed in Beem capsules of capacity 0.25 ml and a capsule was applied to the apical cut surface of an excised node, the whole being immersed in nutrient solution with or without sucrose. Capsules were renewed every 3 days.

Shoot apices, nodes comparable to those used in the experiments, and experimentally treated nodes were fixed in Craf III (Sass[41]) or formalin-acetic-alcohol (Johansen[18]), dehydrated, embedded in Paraplast and sectioned serially at 8 or 10 μ. Sections were stained with safranin, orange G, tannic acid and iron alum

(Sharman[43]) or with 0.1% aqueous toluidine blue.

In some experiments excised nodes were grown for a variable number of days in control medium, and then transferred for 24 hours to medium containing 5 μc/ml ^{3}H-thymidine (Schwarz Bioresearch, specific activity 11.2 c/mmole). Nodes were washed in cold thymidine at approximately 100 times the molarity of the radioactive solution for 30 minutes, fixed in FAA, embedded and sectioned as above. Slides were coated with two parts Ilford K2 emulsion dissolved in 1 part water and maintained at approximately 40°C. Slides were developed for 10 minutes in Kodak D 19 developer, washed in acidified water, fixed, and washed in running water for 30-60 minutes. Sections were stained with freshly made aqueous toluidine blue O (Boyd[2], slightly modified).

RESULTS

Sections showed that as in *Alternanthera philoxeroides* (Amaranthaceae), which has a similar pattern of larger and smaller buds (Cutter[7,8]), opposite bud primordia were of unequal size from the time of their inception, usually in the axils of the third pair of leaf primordia (Fig. 2A). At older nodes, such as those used in the experiments, the difference in size and stage of development was quite marked (Fig. 2B).

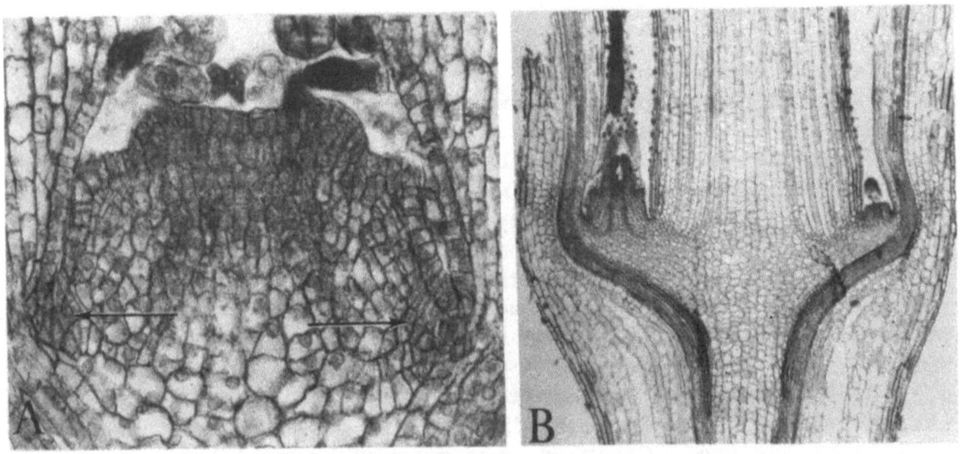

Figure 2. A, L.S. shoot apex of *Hygrophila* showing the inception of unequal bud primordia (arrowed) in the axils of the third pair of leaf primordia. x 375. B, L.S. node comparable to those used in the experiments, showing the unequal axillary buds. x 35.

Experiments on Apical Dominance

The effects of decapitation and of cytokinins on apical dominance were examined in various experiments, two of which will be described briefly. In one experiment shoots of Hygrophila with 4 pairs of expanded leaves were placed in control medium or in medium containing 1 mg/l kinetin, either intact or decapitated; in the latter case, only 3 nodes remained. There were 9 replicates in each treatment. The results after 9 days are presented in Table I.

It appears from these results that in Hygrophila, as in many other plants, buds at the uppermost remaining nodes grow out following decapitation; the shoot which developed at the uppermost node was larger, grew vertically and rapidly assumed dominance. In intact shoots in the presence of kinetin, one bud also grew out at most nodes, indicating that apical dominance had been broken, but the pattern was rather different in that the greatest number of buds grew out at the second node from the base. Thus one bud, the larger or + bud, grows out following these treatments. In shoots to which both treatments were applied, their effects appeared to be additive, in that the larger bud grew out at virtually every node, and both buds grew out at some nodes. Seventeen days after the beginning of the treatment the uppermost two nodes had 2 buds growing out in nearly every case.

TABLE I. EFFECT OF DECAPITATION AND CYTOKININ ON OUTGROWTH OF BUDS

Treatment	No. of shoots with one emerged bud at the node indicated; nodes numbered acropetally. Figures in parentheses indicate the no. of nodes with a second emerged bud.			
Node No.	1	2	3	4
Intact, control medium	0	1	0	0
Decapitated, control medium	0	7	9	
Intact, 1 mg/l kinetin	5	8	6	1
Decapitated, 1 mg/l kinetin	9 (1)	9 (2)	8 (4)	

Since the smaller or - bud does not grow out in response to either treatment alone, but requires both, these results suggest that the - bud is subject to correlative effects other than, or additional to, apical dominance. The results also indicate that in Hygrophila the larger bud can be released from apical dominance either by decapitation or by treatment with a cytokinin, but suggest that these treatments do not have identical effects since the buds at various nodes along the shoot respond differently to them. The differential response to cytokinins of buds situated at different nodes was also shown in another experiment, in which 9 shoots having 6 pairs of expanded leaves were placed in control medium, and 9 comparable shoots in medium containing 0.05 mg/l benzyladenine (BA). After 27 days, no buds had grown out in the control shoots, and one bud had grown out at several nodes of the BA-treated shoots, in the pattern indicated in Fig. 3. The pattern of bud outgrowth shown in Fig. 3 was established earlier, at least by 12 days after the beginning of the experiment.

These results indicate that buds at certain nodes along the shoot, and thus presumably at a certain stage of development or in a certain physiological state, respond preferentially to cytokinin applied to the whole shoot.

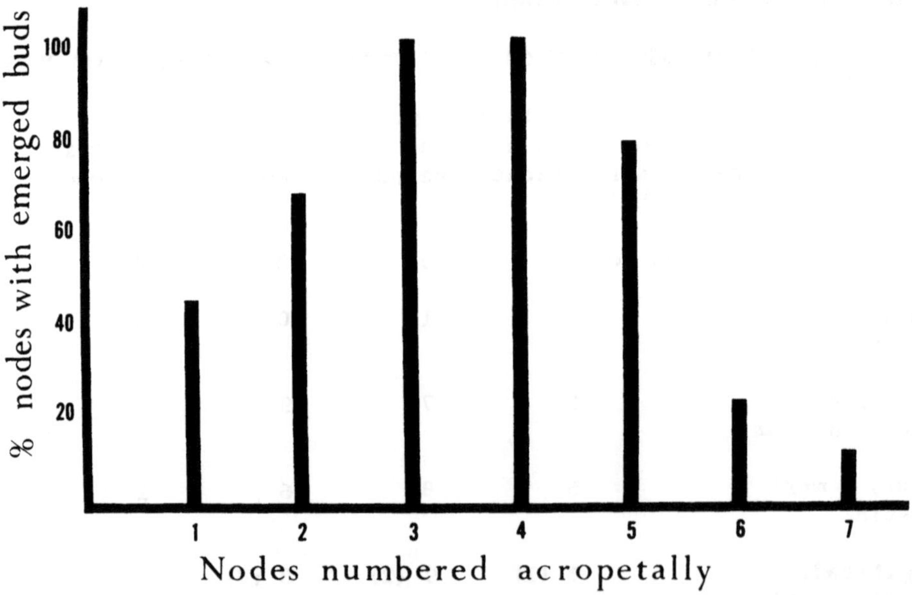

Figure 3. The percentage of larger or + buds which had grown out at each node 27 days after shoots were placed in medium with 0.05 mg/l benzyladenine. In control shoots, no buds grew out.

Correlative Effects between Buds

Since the material was maintained in aseptic culture, it was possible to grow individual excised nodes, and this technique was used to investigate further the correlative effects operative on the - buds. In 32 out of 34 excised nodes one bud, the larger, grew out, thus confirming the result obtained from decapitating shoots. Other nodes were split in half longitudinally and the two halves cultured. In 40 out of 45 such split nodes, both buds grew out (Cutter[8]). Thus it appeared that the larger bud inhibited outgrowth of the smaller bud at the same node.

This conclusion is supported by observation of sections of excised nodes fixed at various times after excision. These reveal that, although never macroscopically visible, the smaller bud does initially undergo some growth following excision, and is thus initially released from apical dominance; it is then re-inhibited, presumably by the larger bud. Histoautoradiographs show that the nuclei of both buds incorporate ^{3}H-thymidine (Fig. 4), and thus also support this view.

In further experiments nodes were split and the two halves grafted together again. Initially halves of the same node, thus having a + and a - bud, were grafted together, and these were termed +/- grafts. In grafts involving halves of the same node, the majority (20 out of 23) developed only one shoot; but if a small piece of aluminum foil was inserted at the nodal region at the time of grafting the majority (12 out of 15) developed two

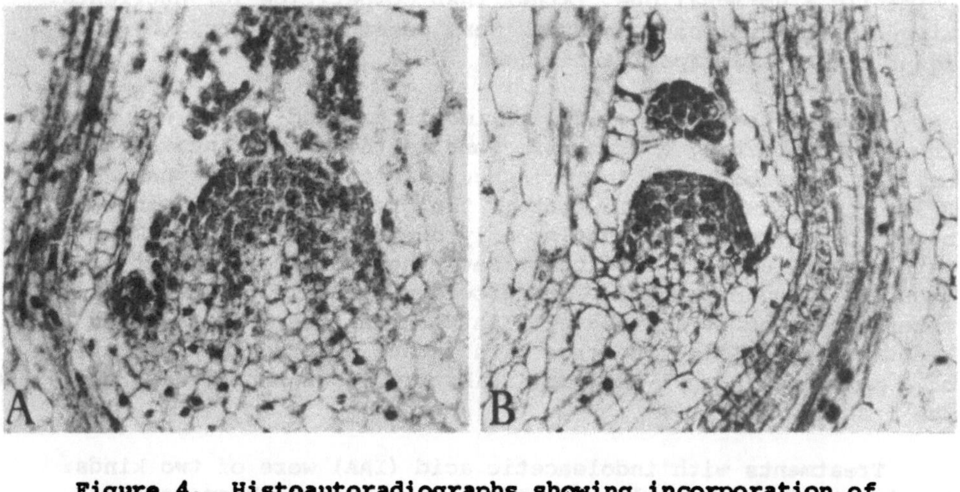

Figure 4. Histoautoradiographs showing incorporation of ^{3}H-thymidine into the nuclei of A, the larger and B, the smaller bud at an excised node grown for 24 hr in control medium and then 24 hr in medium containing 5 μc/ml ^{3}H-thymidine. x 225.

shoots (Cutter and Chiu[9]). Thus the inhibitory effect of the larger bud on the smaller can apparently pass laterally across a graft union, but is prevented by the introduction of an impermeable barrier such as aluminum foil.

In order to determine whether the inhibition would operate between two similar buds, i.e. two + or two - buds, it was necessary to graft together half-nodes which differed in age by two plastochrons. If it is assumed that, say, a + bud occurs on one side of the stem at a particular node, it can be predicted from the helicoidal pattern (Fig. 1) that a similar bud will occur on the opposite side of the stem at a node two plastochrons younger. In grafts between two such half-nodes bearing buds of the same kind (+/+ grafts), the majority (30 out of 42) gave rise to two shoots, the remainder one shoot (Cutter and Chiu[9]). As a control for this experiment, +/- grafts in which the half-nodes differed in age by two plastochrons were also made. Out of 22 such grafts, 11 yielded one shoot and 11 yielded two shoots. The results thus differed from those from +/- grafts of the same node.

Observations of sectioned material, and further experiments conducted for the purpose, indicated that an age differential between the half-nodes, reflected in some specimens but not others by an evident difference in stage of differentiation of the stem, affected the results (Cutter and Chiu[9]). Setting aside this effect, the results seem to show that the inhibitory effect which operates between a + and a - bud is not operative, or greatly lessened, between two + or two - buds. It seemed likely that this effect was primarily a hormonal one, rather than competition for nutrients, since the nutritional conditions were carefully controlled and identical in all experiments.

In a further series of experiments excised nodes were placed in media containing a hormone. Some of these experiments will be briefly described. In the presence of either 1 mg/l kinetin or 0.1 mg/l BA, two buds grew out in the great majority of instances, nodes in control medium having only one shoot (Fig. 5). Treatment with a cytokinin can thus counteract the inhibitory effect of the + on the - bud at a node. Bud growth was much more normal and extensive in nodes treated with BA than in those treated with kinetin, as was also the case with intact shoots. Concentrations of 3 mg/l and 10 mg/l BA were inhibitory to growth of both the - and the + buds.

Treatments with indoleacetic acid (IAA) were of two kinds: treatment of the apical cut surface with plastic Beem capsules containing 1 or 10 mg/l IAA in 0.8% or 1% agar, and immersion of the whole in control medium, or treatment with capsules containing plain agar and immersion in medium containing no, 1 or 10 mg/l IAA. The results of one such experiment, in which sucrose

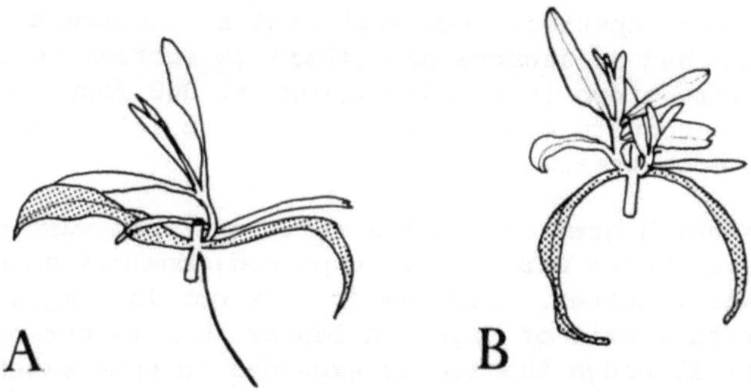

Figure 5. Excised nodes grown in A, control medium, and B, medium with 0.1 mg/l BA for 19 days. One bud has grown out in A, two in B. (Drawn from a photograph; leaves of original node stippled). x 1.

was omitted from the medium but the cultures were maintained in a growth chamber with a light intensity of 2000 foot-candles, are presented in Table II. Nodes of comparable age were used; there were 10 replicates.

Thus IAA was inhibitory to bud outgrowth, especially if supplied in the medium; in this case, of course, greater amounts of the auxin would be continuously available. Those buds which did grow out in medium containing IAA were extremely small.

TABLE II. EFFECT OF IAA ON OUTGROWTH OF THE LARGER BUD IN ISOLATED NODES

Treatment	No. of nodes with 1 bud growing out					
	3 days	7 days	10 days	13 days	16 days	20 days
Control	10	10	10	10	10	10
1 mg/l IAA in capsule	6	10	10	10	10	10
10 mg/l IAA in capsule	0	3	7	7	10	10
1 mg/l IAA in medium	1	4	4	4	6	8
10 mg/l IAA in medium	0	0	2	2	2	2

In a further experiment somewhat similar treatments were given, but some had no sucrose and others 2% sucrose in the medium. The cultures were grown in a culture room at 360 foot-candles light intensity. The results are presented in Table III; there were 10 replicates in each treatment.

The buds which grew out in IAA solutions were very small, with 1 pair of leaves which were hardly expanded; control shoots had 3 pairs of expanded leaves. Buds which grew out in 3 mg/l BA were also small, with 1 pair of expanded leaves, but rather more normal in appearance; although the leaves expanded to some extent the buds did not elongate (Fig. 6). The lack of sucrose had no major effect on control nodes, except that the number of pairs of expanded leaves was rather lower. Absence of sucrose, however, enhanced the inhibitory effects of the hormones. Thus the hormonal and nutritional effects seemed to interact and essentially to be cumulative. Further experiments with varied nutrition seem desirable.

DISCUSSION

The complex problem of apical dominance, and other correlative effects in the shoot, might be more easily understood if critical attention were paid to the morphogenesis of the released buds. From a consideration of the literature it appears possible that in any one species certain factors only affect, or predominantly affect, buds in a particular phase of ontogeny, or at least affect them differentially according to stage of development. In *Linum*, for

TABLE III. EFFECT OF SUCROSE AND HORMONES ON OUTGROWTH OF BUDS IN ISOLATED NODES

Treatment	No. of nodes with 1 emerged bud; figures in parentheses indicate no. of nodes with a second emerged bud			
	2% Sucrose		No Sucrose	
	12 days	18 days	12 days	18 days
Control	10	10	10	10
10 mg/l IAA	5	6	0	0
3 mg/l BA	10 (5)	10 (7)	3	3 (1)
10 mg/l IAA + 3 mg/l BA	0	0	0	0

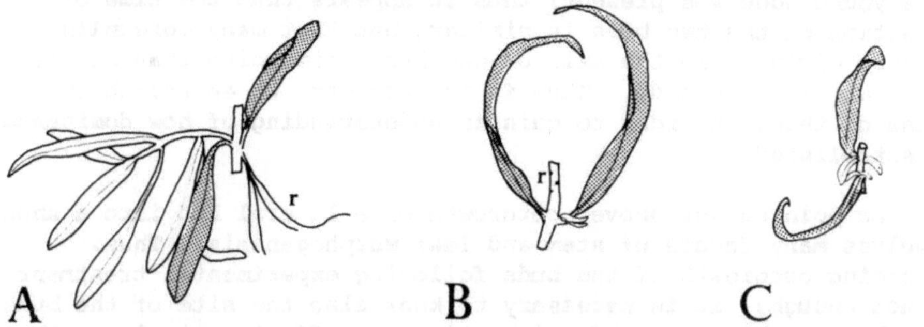

Figure 6. Excised nodes grown in the following media, all with 2% sucrose, for 18 days: A, control; B, 10 mg/l IAA; C, 3 mg/l BA. One bud has grown out in A and B, but is strongly inhibited in presence of IAA; two small buds have grown out in BA (C). (Drawn from a photograph; leaves of original node stippled; r, roots). x 1.

instance, a difference in response to nutritional treatments was noted according to the position of buds along the stem (Gregory and Veale[14]). Phillips[34] made the interesting observation that gibberellic acid (GA) was antagonistic to the effect of IAA on bud outgrowth when applied to the fully elongated second internode of decapitated bean plants, but that if the internode was still elongating at the time of application GA enhanced the inhibitory effect of IAA. This may well mean that the interacting effects of these hormones differ according to the stage of development of the buds themselves. In shoots of Hygrophila treated with a cytokinin, as reported above, the buds at certain nodes along the shoot are more responsive than others. It remains to be determined by sectioning each node how extensively the buds actually differ in stage of development; in species with total apical dominance, like Hygrophila, they probably differ much less than in species with partial dominance. A second possible explanation of the observations is that a substance derived from the shoot apex interacts with the cytokinin and reaches a threshold value at certain positions along the stem.

In Hygrophila, and certain other species with unequal buds, the two buds at the same node differ in size and stage of development. The question of what is the physiological basis of the dominance of the larger over the smaller bud is an interesting one. In the case of apical dominance, it has been suggested that the fact that the main apex was formed before the buds is important (McIntyre[23]). In Asclepias, which has unequal buds of different morphogenetic potentialities, the stronger of the two buds is said to be formed much earlier than the weaker (Nolan[29]). Examination of sectioned shoot apices of Hygrophila and Alternanthera, however, did not unequivocally reveal any in which only one bud primordium at a young node was present; thus it appears that the time of inception of the two buds is similar, but that many more cells undergo division in the axil of one leaf primordium than of the other at the same node. Thus it is necessary to establish the cause of this, in order to gain an understanding of how dominance is established.

As pointed out above, outgrowth of a lateral bud into a shoot involves many facets of stem and leaf morphogenesis. Thus, recording outgrowth of the buds following experimental treatment is not enough: it is necessary to know also the site of the buds and whether the agent under investigation affects apical growth, leaf formation, leaf expansion, internodal extension, etc., or all of these, in the buds. For example, analysis of the effects of splitting the uppermost node in decapitated plants of *Alternanthera philoxeroides*, a species with a pattern of unequal buds similar to that of *Hygrophila*, or of excising the larger bud, showed that not only the height of the smaller shoots, but also the number of pairs of expanded leaves and the average internode length were enhanced in those released from the dominance of the larger bud. Thus growth in both the apical (number of pairs of leaf primordia formed) and subapical (internodal extension) regions of the bud must be affected by the dominance of the larger bud. However, it is known in other dicotyledons that young leaves are apparently sources of the factor stimulating elongation of the internode below (Jones and Phillips[19]; Wetmore and Garrison[53]). Hence morphogenesis of a lateral bud released from inhibition is probably essentially an epigenetic phenomenon, triggered by a stimulus to growth of the shoot apical meristem itself. Where growth has already been initiated, however, as in lateral shoots of species with partial apical dominance, other factors may assume greater relative importance. For example, both IAA and GA can stimulate growth of buds already released from apical dominance, but cannot achieve release (Sachs and Thimann[40]).

The need for observation of bud morphogenesis in studies of correlative inhibition is also emphasized by the fact that unless sectioned material is examined, a temporary release from apical dominance may go unremarked. This is shown clearly by the results with excised nodes of *Hygrophila* described above.

In reaching a correct interpretation of the phenomenon of correlative inhibition a knowledge of the precise sequence of events, for example bud outgrowth and the differentiation of the bud vascular supply, is clearly critical. In the discussion of some of the literature pertaining to apical dominance which follows, particular attention will be paid to the sequence of events in the morphogenesis of released buds, wherever this has been reported.

Auxin appears to exert its inhibitory effects in early stages of bud morphogenesis. In the ferns Matteuccia struthiopteris and Onoclea sensibilis detached meristems comprising a superficial group of meristematic cells occur at particular loci on the rhizome but do not grow out. Following decapitation of the rhizome, however, cell division occurs in the meristem, a distinctive shoot apical cell is delimited and leaf primordia are formed (Wardlaw[49,50]). In this instance apical dominance completely prevents leaf formation and the buds are extremely inhibited. Application of auxin to decapitated rhizomes maintains this inhibition (Wardlaw[51]). Further evidence that auxin acts on early stages of bud morphogenesis, and indeed on the bud meristem itself, comes from the work of Naylor[27]. In shoots of Tradescantia with complete apical dominance, he showed that no mitosis occurred in the cells of the terminal region of the apex of lateral buds and that interphase nuclei in this region were in the 2C condition. Three days after decapitation of the shoot most of these nuclei were in the 4C condition, and mitosis ensued. If an auxin was applied to the decapitated shoot the bud nuclei did not attain the 4C condition. Thus auxin apparently acted by inhibiting DNA synthesis in these buds. In general, however, little is known of the mode of action at the cellular level of auxin or of other hormones involved in apical dominance.

To achieve its inhibitory effect, auxin must be applied very soon after decapitation of the shoot and must be continuously available (Panigrahi and Audus[30]; Thimann, Sachs and Mathur[47]; Wickson and Thimann[56]). This probably explains why in the experiments with Hygrophila application of IAA in the nutrient solution was more effective than application in apically-applied Beem capsules, since these may not have been renewed sufficiently frequently. Even buds growing in solutions containing IAA, however, ultimately grow out to some extent (Tables II and III); similar effects have been reported also in peas (Jacobs and Case[16]; Wickson and Thimann[56]). Thus the inhibitory effect of auxin is apparently somewhat transitory; it is also not operative in growing apices (Sachs and Thimann[40]).

In Coleus, which has partial apical dominance, applied auxin apparently had no inhibitory effect on the lateral shoots of decapitated plants, and it was considered that its mechanism of apical dominance might differ from that of other plants (Jacobs, Danielson, Hurst and Adams[17]). (It may be noted, incidentally, that in these experiments applied auxin consistently stimulated the growth of buds in the axils of the third pair of leaves, which were presumably at a particular stage of development). More recently, it has been shown that under certain conditions auxin does indeed inhibit growth of the axillary buds of Coleus, but is more effective when applied in aqueous solution than in lanolin (Thimann, Sachs and Mathur[47]). Moreover, apical dominance is much stronger, and buds are more sensitive to inhibition by auxin, in plants grown under low light intensities or poor nutritional conditions.

There is thus good evidence that auxin inhibits outgrowth of lateral buds in many species, at least for a while, and that it is most effective on buds which are in early stages of morphogenesis. The balance between auxin and cytokinin at the bud locus is considered to be important in controlling apical dominance, and indeed the discovery that kinetin applied directly to the apex of an inhibited bud could release it from apical dominance was an important step in the study of this phenomenon (Sachs and Thimann[39,40]; Wickson and Thimann[56]). This release was complete only for the first few days, subsequent growth being less than in decapitated plants (Panigrahi and Audus[30]; Sachs and Thimann[39]). However, in Hygrophila, and in other experiments with Alternanthera (Cutter[7]), the cytokinin benzyladenine gave much more normal bud growth than kinetin; in tobacco, also, BA stimulated bud growth to the same extent as decapitation (Schaeffer and Sharpe[42]). In Hygrophila, however, the results suggest that the effects on bud outgrowth of decapitation and cytokinin treatment are not identical, since buds at different nodes were preferentially stimulated by the two treatments. In species with determinate shoots, the balance between the levels of gibberellin and cytokinin is considered important in controlling bud outgrowth (Fisher[11]).

In Hygrophila, treatment with a cytokinin released buds either from the inhibitory effect of the main apex, or from that of the larger bud at the same node, suggesting that these two correlative effects are physiologically similar. Sachs[36] also found that application of kinetin to the smaller cotyledonary shoots of decapitated pea seedlings released them from inhibitory effects of the larger shoots. It has been argued (McIntyre[24]) that this effect is consistent with the hypothesis that apical--or bud--dominance depends on competition for nutrients, on the grounds that localized application of kinetin leads to mobilization of nitrogenous materials to the site of application. However, in the experiments on Hygrophila the cytokinins were not applied locally, but in the medium in which the shoots were immersed, and it seems unlikely that this explanation would hold.

Since cytokinin is involved in the actual release from apical dominance it, too, evidently acts on early stages of bud morphogenesis. In mosses, at least, cytokinin does not act as a trigger on bud formation and development but has to be present throughout a critical period (Brandes and Kende[1]). Whether this is true also in higher plants remains to be determined.

A number of workers have suggested that apical dominance, and especially auxin from the shoot tip, exert their effects by inhibiting the differentiation of vascular tissue between the axillary buds and the main stem (Gregory and Veale[14]; Panigrahi and Audus[30]; Sachs[38]; Sorokin and Thimann[46]). Although the absence of good vascular connections to the root system would clearly eventually limit the growth of the lateral shoots, a critical consideration of the sequence of these events--where this is reported--suggests that a differentiated vascular system is not necessary for the initial outgrowth of the buds. Moreover, it should be borne in mind that the main shoot apex itself, and many of the developing leaf primordia around it, are devoid of differentiated vascular tissues but nevertheless undergo continuous meristematic growth; in plants of Linum with high systems of phyllotaxis, for example, as many as 32 leaf primordia were devoid of mature phloem elements, and the 48 youngest primordia had no mature xylem (Esau[10]). There is much evidence, also, that growing buds induce the differentiation of vascular tissues (Camus[3]; Wetmore and Rier[54]; Wetmore and Sorokin[55]), so that it may be questioned whether the vascular tissue is necessary for, or actually results from, bud outgrowth.

Although the observation that in excised stem segments of pea auxin somewhat inhibited, and kinetin promoted, both bud outgrowth and contact between the xylem of lateral buds and that of the stem, seems to support the view that bud inhibition may be due to inadequate contact of the bud xylem (Sorokin and Thimann[46]), consideration of the reported sequence of events renders this less plausible. The buds grew quite considerably (about 20% in length) during the first 24 hours; differentiated xylem remained discontinuous, however, until 50-56 hours after excision, and was fully continuous only after 72 hours (Sorokin and Thimann[46]). In another study with peas, lateral buds showed marked stimulation of growth 8 hours after decapitation of the shoot; growth probably began within 4 hours. However, there was no sign of differentiated xylem 1 day after decapitation, and sieve plates were not detected until 4 days (Wardlaw and Mortimer[52]). In this instance it was concluded that this sequence suggested that apical dominance did not act through inadequate differentiation of the bud vascular system. In Helianthus, also, axillary buds grew markedly within 2 days of decapitation of the shoot, before vascular connections existed (Marr and Blaser[22]), and in the experiments with Hygrophila bud outgrowth began while only procambium was present in the bud trace.

By ingenious experiments Sachs[37,38] has shown that either applied auxin or active parts of the shoot, such as leaves, can inhibit the contact between strands of xylem; he showed also that naturally released buds on intact plants had a relatively independent vascular system that did not connect directly with the stem vascular system, whereas after artificial release, such as decapitation, contact was established. The vascular system was studied when the buds were already 1-3 cm long. Sachs[38] considers that the control of bud growth through the control of vascular differentiation is established, but that other mechanisms of apical dominance may also exist. However, on balance it seems likely to the present writer that the bud vascular system does not control the initial growth of the bud, although clearly continued lack of vascular connections would ultimately limit growth.

Various workers have claimed that apical dominance can be wholly or partially explained in terms of competition for nutrients (Gregory and Veale[14]; McIntyre[24,25]) or for a limited water supply (McIntyre[25,26]). The inability of the buds to compete successfully for nutrients is sometimes attributed to the supposed inadequacy of the bud vascular system (Gregory and Veale[14]; Panigrahi and Audus[30]). There can be little doubt that a shortage of nutrients would inhibit lateral shoot growth at some stage, but whether it is the sole factor maintaining total apical dominance is more doubtful, though under some conditions it may be a limiting factor. McIntyre[24] found that in decapitated flax seedlings with two shoots in the axils of the cotyledons the inhibited shoot could be released by increasing the supply of nitrogen or phosphorus. He therefore argued that the inhibitory effect of one shoot on the other was due primarily to competition for a limited nutrient supply. In decapitated shoots such as these, both buds were initially released from apical dominance; McIntyre[24] stated that the axillary bud primordia of flax seemed to be of equal size, and the grafting experiments on Hygrophila outlined above seem to show (1) that the inhibitory effect of one bud on another is hormonal in nature, and (2) that this effect is not operative between equal buds. The latter conclusion is supported by observation of decussate plants with equal buds, such as Acer, in which two equal shoots usually grow out. Thus in flax the hormonal conditions may have been such that both buds could grow, but under conditions of limited nutrition competition between them became important. McIntyre[24] found an increase in the nitrogen content of the inhibited shoot 12 hours after excision of the dominant shoot, but this might well be a result, rather than a cause, of its growth. Phillips[31] analyzed the total content of N, P and K per unit dry weight of the apical 5 mm of axillary buds of decapitated bean plants, and found that it was greater in those inhibited by applied IAA than in the growing buds of control plants. It thus seemed likely that the availability of inorganic nutrients was not a limiting factor in the early outgrowth of these buds. In broad bean, the movement of uracil-2-C^{14} from the cotyledons into the lateral buds was greatly enhanced by decapitation, and it was suggested that the accelerated movement of nutrients may actually precede and be a cause of bud growth (Panigrahi and Audus[30]). However, this substance may perhaps be a rather specific requirement in the released buds, in view of the relationship between cytokinin-induced growth and DNA-dependent RNA synthesis (Khan[20]).

The nutritional status of plants may affect the action of hormones on bud outgrowth (Catalano and Hill[4]; Phillips[33]). For example, in Solanum the lower axillary buds of intact plants were released from complete inhibition by high applications of potassium to the soil. These buds did not respond to kinetin, but the completely inhibited buds of low potassium plants did, suggesting that high levels of potassium might affect the level of endogenous cytokinins (Wakhloo[48]). Since water stress can result in the inactivation of cytokinins (Itai and Vaadia[15]), outgrowth of lateral buds of pea resulting from a reduction in water stress (McIntyre[26]) might also be explicable in terms of a higher level of available cytokinin. In Hygrophila, also, there was an interaction between nutrition and hormones, since the inhibitory effects of both IAA and high concentrations of BA were enhanced in the absence of sucrose. It seems unlikely in this instance, however, that better carbohydrate nutrition stimulated cytokinin synthesis, since bud growth in already inhibitory concentrations of cytokinin was improved in the presence of sucrose.

In general, indeed, it seems unnecessary either to interpret the stimulatory effects of nutrition on bud growth in terms of hormones, or the effects of hormones in terms of the mobilization of nutrients. Both correct hormonal conditions and adequate nutrients are no doubt required for bud outgrowth, and presumably either of these, and conceivably also other factors, could be limiting at a particular stage of morphogenesis.

In attempting to interpret apical dominance it is important to think not only of the physiological and morphogenetic events taking place in the cells and organs of the buds released from dominance, but also of those, presumably converse, changes occurring when dominance is being imposed. It seems certain that more precise observations of the sequence of events in lateral bud morphogenesis, and a study of the effects of various factors on the cells of the bud apical meristem, would contribute substantially to an understanding of correlative inhibition. Apparently IAA applied to the decapitated stem inhibits DNA synthesis in lateral buds (Naylor[27]), and BA applied to the buds promotes DNA synthesis (Schaeffer and Sharpe[42]). No doubt any interpretation along these lines would be greatly over-simplified, but further work of this kind, involving also a study of various regions of the bud meristem, would surely be fruitful.

REFERENCES

1. Brandes, H. and Kende, H. (1968) Studies on cytokinin-controlled bud formation in moss protonemata. Plant Physiol. 43:827-837.

2. Boyd, G. A. (1955) Autoradiography in Biology and Medicine. Academic Press, New York.

3. Camus, G. (1943) Sur le greffage de bourgeons d'Endive, sur des fragments de tissus cultives in vitro. C. R. Acad. Sci., Paris 137:184-185.

4. Catalano, M. and Hill, T. A. (1969) Interaction between gibberellic acid and kinetin in overcoming apical dominance, natural and induced by IAA, in tomato (Lycopersicum esculentum Mill. cultivar Potentate). Nature 222:985-986.

5. Champagnat, P. (1958) Ramification a regime rhythmique et anisophyllie chez les vegetaux superieurs. Lilloa 16:161-191.

6. Cutter, E. G. (1966) Patterns of organogenesis in the shoot. In Trends in Plant Morphogenesis, ed. E. G. Cutter, et al.: 220-234. Longmans, London.

7. Cutter, E. G. (1967) Morphogenesis and developmental potentialities of unequal buds. Phytomorphology 17:437-445.

8. Cutter, E. G. (1972) Regulation of branching in species with unequal lateral buds. Ann. Bot. (In press).

9. Cutter, E. G. and Chiu, H. (1972) Grafting experiments on correlative effects between lateral buds. Ann. Bot. (In press).

10. Esau, K. (1943) Vascular differentiation in the vegetative shoot of Linum. II. The first phloem and xylem. Amer. J. Bot. 30:248-255.

11. Fisher, J. B. (1971) Control of bud inhibition in Cyperus. Planta 97:257-268.

12. Gautheret, R. J. (1959) La Culture des Tissus Vegetaux. Masson et Cie., Paris.

13. Goebel, K. (1928) Organographie der Pflanzen. Vol. 1. Gustav Fischer, Jena.

14. Gregory, F. G. and Veale, J. A. (1957) A reassessment of the problem of apical dominance. Symp. Soc. Exp. Biol. 11:1-20.

15. Itai, C. and Vaadia, Y. (1971) Cytokinin activity in water-stressed shoots. Plant Physiol. 47:87-90.

16. Jacobs, W. P. and Case, D. B. (1965) Auxin transport, gibberellin, and apical dominance. Science 148:1729-1731.

17. Jacobs, W. P., Danielson, J., Hurst, V. and Adams, P. (1959) What substance normally controls a given biological process? II. The relation of auxin to apical dominance. Dev. Biol. 1:534-554.

18. Johansen, D. A. (1940) Plant Microtechnique. McGraw-Hill, New York.

19. Jones, R. L. and Phillips, I. D. J. (1966) Organs of gibberellin synthesis in light-grown sunflower plants. Plant Physiol. 41:1381-1386.

20. Khan, A. A. (1971) Cytokinins: permissive role in seed germination. Science 171:853-859.

21. Loiseau, J.-E. (1969) La Phyllotaxie. Monograph 4. Masson et Cie., Paris.

22. Marr, C. and Blaser, H. W. (1967) The origin and development of induced branches in Helianthus. Amer. J. Bot. 54:498-504.

23. McIntyre, G. I. (1964) Mechanism of apical dominance in plants. Nature 203: 1190-1191.

24. McIntyre, G. I. (1968) Nutritional control of the correlative inhibition between lateral shoots in the flax seedling (Linum usitatissimum). Can. J. Bot. 46:147-155.

25. McIntyre, G. I. (1971a) Apical dominance in the rhizome of Agropyron repens. Some factors affecting the degree of dominance in isolated rhizomes. Can. J. Bot. 49:99-109.

26. McIntyre, G. I. (1971b) Water stress and apical dominance in Pisum sativum. Nature New Biol. 230:87-88.

27. Naylor, J. M. (1958) Control of nuclear processes by auxin in axillary buds of Tradescantia paludosa. Can. J. Bot. 36: 211-232.

28. Nitsch, J. P. (1951) Growth and development in vitro of excised ovaries. Amer. J. Bot. 38:566-577.

29. Nolan, J. R. (1969) Bifurcation of the stem apex in Asclepias syriaca. Amer. J. Bot. 56:603-609.

30. Panigrahi, B. M. and Audus, L. J. (1966) Apical dominance in Vicia faba. Ann. Bot. 30:457-473.

31. Phillips, I. D. J. (1968) Nitrogen, phosphorus and potassium distribution in relation to apical dominance in dwarf bean (Phaseolus vulgaris, c.v. Canadian Wonder). J. Exp. Bot. 19:617-627.

32. Phillips, I. D. J. (1969a) Apical dominance. In The Physiology of Plant Growth and Development, ed. M. B. Wilkins: 165-202. McGraw-Hill, London.

33. Phillips, I. D. J. (1969b) Auxin-gibberellin interaction in apical dominance: experiments with tall and dwarf varieties of pea and bean. Planta 86:315-323.

34. Phillips, I. D. J. (1971) Effect of relative hormone concentration on auxin-gibberellin interaction in correlative inhibition of axillary buds. Planta 96:27-34.

35. Raciborski, M. (1901) Über die Verzweigung. Annls Jard. Bot. Buitenz. 17:1-67.

36. Sachs, T. (1966) Senescence of inhibited shoots of peas and apical dominance. Ann. Bot. 30:447-456.

37. Sachs, T. (1968) On the determination of the pattern of vascular tissue in peas. Ann. Bot. 32:781-790.

38. Sachs, T. (1970) A control of bud growth by vascular tissue differentiation. Israel J. Bot. 19:484-498.

39. Sachs, T. and Thimann, K. V. (1964) Release of lateral buds from apical dominance. Nature 201:939-940.

40. Sachs, T. and Thimann, K. V. (1967) The role of auxins and cytokinins in the release of buds from dominance. Amer. J. Bot. 54:136-144.

41. Sass, J. E. (1958) Botanical Microtechnique. 3rd ed. Iowa State Univ. Press, Ames, Iowa.

42. Schaeffer, G. W. and Sharpe, F. T., Jr. (1969) Release of axillary bud inhibition with benzyladenine in tobacco. Bot. Gaz. 130:107-110.

43. Sharman, B. C. (1943) Tannic acid and iron alum with safranin and orange G in studies of the shoot apex. Stain Technol. 18:105-111.

44. Snow, R. (1931) Experiments on growth and inhibition. II. New phenomena of inhibition. Proc. Roy. Soc. B 108: 305-316.

45. Snow, R. (1937) On the nature of correlative inhibition. New Phytol 36:283-300.

46. Sorokin, H. P. and Thimann, K. V. (1964) The histological basis for inhibition of axillary buds in Pisum sativum and the effects of auxins and kinetin on xylem development. Protoplasma 59:326-350.

47. Thimann, K. V., Sachs, T. and Mathur, K. N. (1971) The mechanism of apical dominance in Coleus. Physiol. Plant. 24:68-72.

48. Wakhloo, J. L. (1970) Role of mineral nutrients and growth regulators in the apical dominance in Solanum sisymbrifolium. Planta 91:190-194.

49. Wardlaw, C. W. (1943a) Experimental and analytical studies of pteridophytes. I. Preliminary observations on the development of buds on the rhizome of the ostrich fern (Matteuccia struthiopteris Tod.). Ann. Bot. 7:171-184.

50. Wardlaw, C. W. (1943b) Experimental and analytical studies of pteridophytes. II. Experimental observations on the development of buds in Onoclea sensibilis and in species of Dryopteris. Ann. Bot. 7:357-377.

51. Wardlaw, C. W. (1946) Experimental and analytical studies of pteridophytes. VIII. Further observations on bud development in Matteuccia struthiopteris, Onoclea sensibilis, and species of Dryopteris. Ann. Bot. 9:117-132.

52. Wardlaw, I. F. and Mortimer, D. C. (1970) Carbohydrate movement in pea plants in relation to axillary bud growth and vascular development. Can. J. Bot. 48:299-237.

53. Wetmore, R. H. and Garrison, R. (1966) The morphological ontogeny of the leafy shoot. In Trends in Plant Morphogenesis, ed. E. G. Cutter et al.:187-199. Longmans, London.

54. Wetmore, R. H. and Rier, J. P. (1963) Experimental induction of vascular tissues in callus of angiosperms. Amer. J. Bot. 50:418-430.

55. Wetmore, R. H. and Sorokin, S. (1955) On the differentiation of xylem. J. Arnold Arbor. 36:305-317.

56. Wickson, M. and Thimann, K. V. (1958) The antagonism of auxin and kinetin in apical dominance. Physiol. Plant. 11:62-74.

*The experimental work was supported in part by NSF grants GB-6591 and GB-12905, for which grateful acknowledgement is made. I am also indebted to Mr. H. Chiu, Miss M. J. Mahony, and Mr. L. J. Feldman for technical assistance at various times.

-DISCUSSION-

VAN'T HOF - Calculation of the growth fraction in large and small buds and determining whether growth was exponential or linear may give a hint to the basis for size differences. For example, in the larger buds after mitosis each daughter cell may remain meristematic whereas in the smaller buds only one daughter cell may, on the average, continue to divide. Also, after looking at the bud size gradient along the longitudinal axis one wonders if cell loss is a function of age.

CUTTER - If you look carefully at the size of the small buds along the axis, they all appear to cease growth at a similar stage.

DAVIDSON - Did the two kinds of buds reach their maximum size and stop growing at the same time, or does the larger bud cease to grow first?

CUTTER - No, the other way around.

DAVIDSON - In this correlative inhibition you imply that gradients exist across the stem. Do you really think that auxins, or kinetins can move in this direction?

CUTTER - It certainly appears like this. I prefer this explanation to a nutritional inhibition.

HABER - Transverse gradients need not to be invoked if you think about Leopold's experiments of senescence of opposite leaves (Am. J. Bot., 51:294-298, 1964). When he delays the senescence of one by cytokinin treatments, the senescence of the other is greatly accelerated; so any vigorous organ is likely to rob its neighbor of supplies of nutrients and thereby limit its growth.

BALL - Exactly what do you mean by the terms "large" and "small" buds? Are you referring to the size of the shoot apex, or to the length of the shoot produced, or the number of pairs of leaves? It seems to me the shoot apex was about equal in all of your buds.

CUTTER - The diameter of the shoot apex is greater in the larger buds, and there is a larger number of pairs of leaves; and the meristem that gives rise to the shoot apex is also larger from the start. There are more cells involved.

HABER - Is it a fact that you have a perfectly helicoidal arrangement of large and small buds up the stem?

CUTTER - Yes, but it is not absolutely regular all the way up in this species. Occasionally you get a reversal of the helix.

TORREY - I would like to ask what is the operational part of the bud either as source of suppression or as the site of the sink. Is it the apical dome or the leaves?

CUTTER - I think it's the young leaves that produce suppressing substances.

TORREY - But presumably the difference in size between the apical domes must have some influence?

CUTTER - Yes.

VAN'T HOF - The morphological pattern in many meristems is preprogrammed in the embryo. Do you think this is the case in your system?

CUTTER - I wouldn't think so since all our material has been vegetatively propagated for many generations.

VAN'T HOF - Even so you may still be propagating the program.

KONZAK - There is one point that argues against this, namely that complete regularity of the helicoidal arrangement is not present.

THE SURFACE "HISTOGEN" OF LIVING SHOOT APICES

Ernest A. Ball

University of California, Irvine

Theories to account for the growth of the shoot apex of vascular plants include those of the apical cell, histogens, the tunica-corpus, and the initial ring-quiescent center.

The apical cell theory is oldest, and derives from the emphasis upon the individual cell after the Cell Theory was presented. It was supported by numerous investigators in the 19th century. Prominent among them were Hofmeister[24], Nägeli[37], Koch[26], Schüepp[44], Dingler[16], Korschelt[27], Douliot[17], Pringsheim[39], as reviewed by Foster[20]. It is only of historical interest today, since later investigations present much evidence that the finding of a putative apical cell within a shoot apex was regarded by the early workers as a necessity to support the preconceived notion that one had to be present.

The histogen theory was presented by Hanstein[23] and recognized that an apical cell did not exist in the shoot apex of vascular plants above the evolutionary level of the ferns. He thought that the shoot apex consisted of layers of cells which produced specific parts of the primary plant body. His dermatogen, or surface layer, supposedly produced the epidermis; the next underlying layer or layers, the periblem, he considered to produce the inner tissues of the leaf and the cortex of the stem; the inner non-layered tissue of the shoot apex, the plerome, he thought produced the pith and the precursor tissue of the vascular tissue (xylem and phloem). While this theory is also mainly of historical interest, there is some modern evidence to support it. It fell from favor among scientists because many cases of development were described in which there were apparently no strict predestinations of the three supposed histogens in the origin of tissues in the primary plant body. However, modern work by Satina and Blakeslee[41] on Datura gives evidence that the layers of cells in the shoot apex do have predestinations to produce specific tissues in the primary plant body. Though the strict concept of histogens, viz., dermatogen producing exclusively epidermis, periblem producing only internal leaf tissues and cortex, plerome producing only pith and vascular tissues, was not supported categorically by Satina and Blakeslee, their evidence of specific precursors in the shoot apex for mature tissues of the primary plant body is the first sequential relationship between meristem and mature plant. It introduced the technique of "tracers" in the ontogeny of the plant body from the shoot apex, and is thus of great value in that it makes interpretations necessarily objective.

The tunica-corpus theory was conceived by Schmidt[42] to accommodate the fact that often there is apparently no histogenic relationship between layers of cells in the shoot apex and the mature shoot tissues. He emphasized the lack of predestination of specific regions of the shoot apex and the mature tissues derived from it. The tunica, according to him, consisted of a layer or layers of cells covering the shoot apex; the corpus was the internal non-layered tissues of the dome of meristem. These zones of meristematic tissues were merely positional relationships, and did not connote any predestination of tissues. Indeed, there is no provision for initial cells in the tunica-corpus theory; no suggestion is made as to how either tunica or corpus perpetuate themselves in the shoot apex. While the tunica-corpus theory is useful for purely descriptive purposes, it has no value in morphogenesis which has as its main goal the origin of cells, tissues and organs.

The initial ring-quiescent center theory was set forward by Buvat[9] and was based upon earlier hypotheses of Plantefol[38] and considers the shoot apex to be divided into a central region of non- or seldom-dividing cells, the quiescent center, and a peripheral ring of rapidly dividing cells, the initial ring. The quiescent center is supposed to be carried upward by the divisions of the initial ring which also produces the foliar primordia and the axillary buds. The particulars of this theory are based upon purely observational data from killed, paraffin-embedded apices, and no experiments of the author supports it.

None of these theories account for the origin of cells in the shoot apex, i.e., which part of the apex produces the cells that maintain the "histogens" or are utilized in the formation of the successive foliar primordia. As maintained by Soma and Ball[45], markers of small regions of cells, and later observations of their fate can yield evidence of the directions and locations of formation of cells in the shoot apex. In contrast, Loiseau[31] admitted that the displacement of carbon spots from the center of the apex signified cell formation there, but he denied that this growth subtracted from the role of flank tissues as the main source of cells in the meristem. This method of observation of the living shoot apex has obvious advantages over the traditional one of the study of sections of killed meristems. In the latter, a view is obtained of the status of the meristem, modified by the killing and processing, at only a particular time in its ontogeny. Soma and Ball[45] and Ball[3] demonstrated that the shoot apices of *Lupinus albus*, *Vicia faba* and *Asparagus officinalis* are dynamic systems with continual positional changes of cell populations. The present work was undertaken to utilize more refined technique in the study of growth of the surface of the shoot apex with the view of gaining understanding of its putative histogens.

MATERIALS AND METHODS.--Plants of a relatively slow-growing herb, *Trachymene coerulea* (60 days old) and of a rapidly growing one, *Vicia faba* (14 days old) were used in tracer experiments of areas of the shoot apex. The terminal bud was opened by pushing aside the older leaves and removing some of the intermediate-sized foliar primordia, finally bareing the dome of meristem bordered by the youngest three or four lateral appendages.

Having found previously (Soma and Ball[45]) that carbon spots applied to an apex are left behind in the same way as a killed cell, the present work utilized the former since it can be followed by photography of the surface, while the latter necessitates serial sections and the attendant artefacts. Animal charcoal was mixed with silicone stopcock grease or lanolin to produce a black mixture that could be applied as a discrete spot on various areas of the shoot apex. It had been earlier ascertained that spots of lanolin (Ball[1]) or siicone stopcock compound (Soma and Ball[45]) have no effect on the growth processes. Apparently it was not, however, ascertained by Loiseau[31] whether the albumin glue, butter, or peanut oil used by him to mix with lampblack to mark spots on the shoot apex of *Impatiens roylei* had a physiological effect on the meristem.

Spots were applied by a micromanipulator from the flat tip of a microscalpel. Only the drop of suspended carbon touched the dome of meristem. Great care was exercised to avoid contact of the metal of the tool with the shoot apex. Any apices that were accidentally wounded were discarded. Two orientation systems were followed. In the first, a carbon spot ca. 20um in diameter was placed upon the apparent center of the shoot apex (fig. 1, 2). Recalling the earlier results of Soma and Ball[45], in which such an oriented carbon spot had been left behind asymmetrically by the surface growth of the shoot apex of Lupinus albus, an additional system was devised. This consisted of placing, by the micromanipulator, of two (fig. 3), three (fig. 4), or four (fig. 5) carbon spots around the apparent center of the shoot apex. For *Trachymene coerulea*, which has helical phyllotaxy, thse multiple spots were placed randomly with respect to the youngest visible foliar primordia. *Vicia faba*, which has distichous phyllotaxy, was amenable to more precise orientation.

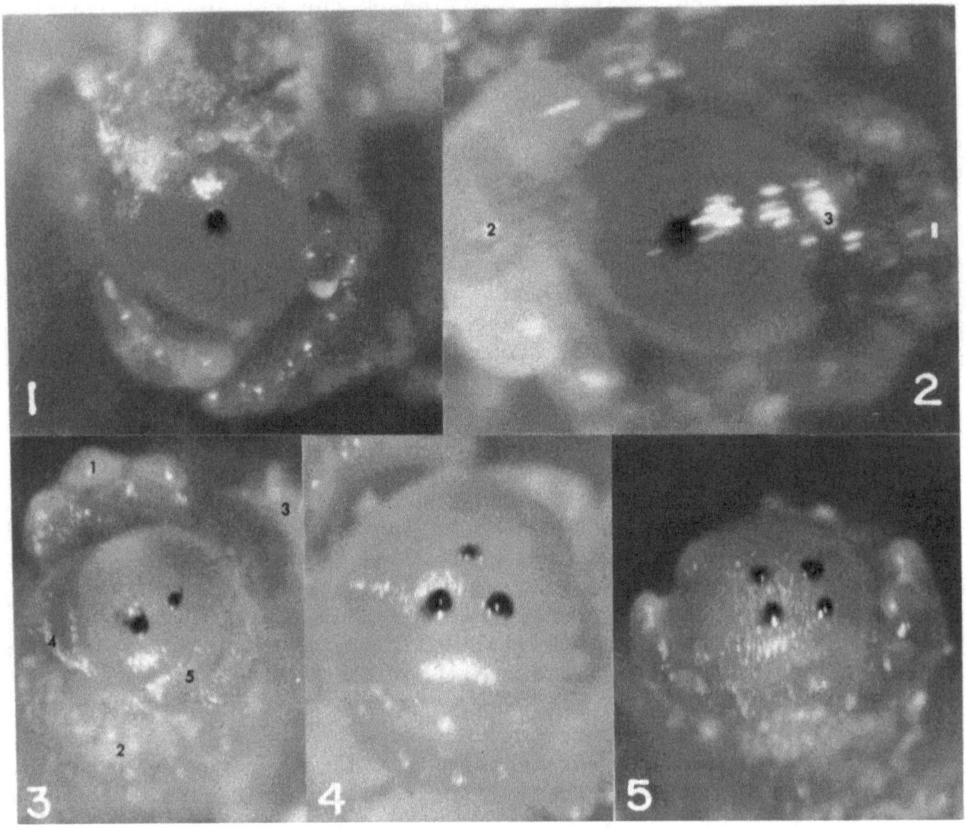

Fig. 1-5.--Fig. 1. Shoot apex of 60 day-old plant of Trachymene immediately after a carbon spot had been placed on its center. 75x. --Fig. 2. Shoot apex of 14 day-old plant of Vicia immediately after a carbon spot had been placed on its center. 150x.--Fig. 3. Shoot apex of 60 day-old plant of Trachymene immediately after two carbon spots had been placed around its presumed center. The last five foliar primordia are numbered in sequence in their helical arrangement. 75x.-- Fig. 4. Shoot apex of 60 day-old plant of Trachymene immediately after three carbon spots had been placed around its presumed center. 75x.-- Fig. 5. Shoot apex of 60 day-old plant of Trachymene immediately after four carbon spots had been placed around its presumed center. 75x.

Here two spots were applied around the putative center of the shoot apex-- either parallel to the "line" of production of the two rows of foliar primordia (fig. 6)or opposite to them (fig. 7).

After the operations were completed, the older leaves were re-bent upward to ensheath the shoot apex and wrapped with a strip of plastic to prevent drying of the meristematic regions.

From one to three weeks after the operation, the leaves were removed from the terminal bud, the stem tip was excised, and under a high-power Greenough dissecting microscope the shoot apex was again laid bare. The stem tip was then affixed to the stage of a Leitz Ultropak microscope on an optical bench. The apex was viewed by a 3.8X Ultropak objective and either a 10X or a 5X ocular. Bellows length of the camera was one meter. Negatives were taken at 150X or 75X on SXX Pan film. Because of the low light intensity and the generally low albedo of the shoot apex, exposures were from 15 to 30 seconds.

The author has based the interpretations of results from these experiments upon the valid assumption that there is no gliding growth of cells or tissues in the shoot apex. Gliding growth probably exists in the vascular cambium where initial cells show extensive tip elongation (Bannan[4]), and intrusive growth occurs by tips of certain primary fibers (Schoch-Bodmer and Huber[43], Esau[18]). Since the few probable cases of gliding growth all involve cells that are conspicuously elongating, and since there are no such cells in the shoot apex, the phenomenon may be logically ruled out for the dome of meristem. The corollary of this assumption is that the cells of the shoot apex do not move over each other. The cells under the surface layer of the shoot apex are left behind without movement from their original positions with respect to the surface layer (cf. Soma and Ball[45]). However, movement of cells within the shoot apex has been assumed by Bergann and Bergann[6] to account for changes in periclinal chimeras, and by Lyndon[32-36] to explain the putative "displacements" during normal growth.

RESULTS.--Trachymene coerulea. The single carbon spot placed at the center of the shoot apex was frequently left behind intact in a lateral position, usually in an axil of a foliar primordium (fig.8), though it was sometimes found in a lateral position circumferentially between two adjacent primordia (fig. 9). These cases illustrate an important principle in the growth of this apex, namely, that a growth center at the position of the future primordium #5 (between #2 and #3) had produced cells that had carried the spot to the other side of the apex. In some cases the intact spot was atop a foliar primordium (fig. 10). Here the spot ended opposite the growth center of the future leaf #6 (between leaves #1 and #4). When the spot was left behind laterally by particularly rapid growth of the shoot apex, a foliar primordium had been formed radially to the outside

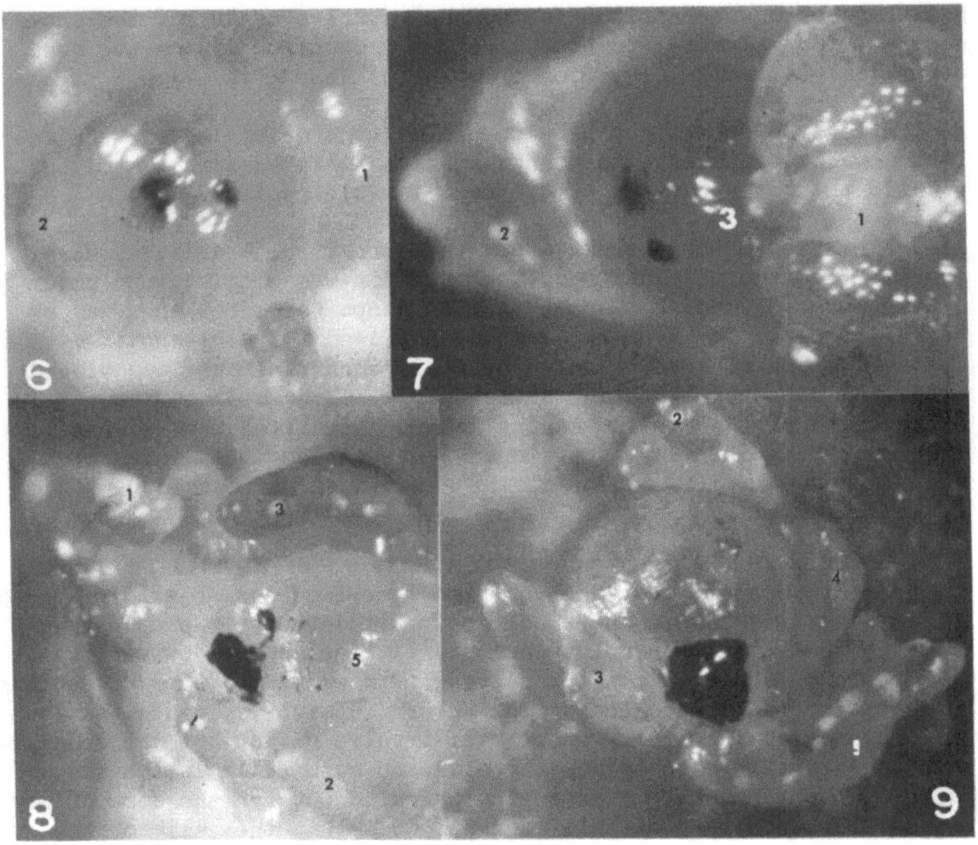

Fig. 6-9.--Fig. 6. Shoot apex of 14 day-old plant of Vicia after two carbon spots had been placed around its presumed center in a line parallel with the two rows of leaves; 2 is the youngest leaf and 1 is the next youngest one, as they appeared in the distichous phyllotaxy. 150x.--Fig. 7. Shoot apex of a Vicia plant 14 days old after two carbon spots had been placed around its presumed center and above the lateral non-leaf producing sides of the meristem. 150x. --Fig. 8. Shoot apex of Trachymene eight days after a carbon spot had been placed upon its center. The carbon spot had been spread somewhat, but the major intact portion had been left behind in the axil of leaf #4. The spot had been left on the opposite side of the apex from the growth center that had produced the newest foliar primordium, #5. 75x.--Fig. 9. Shoot apex of Trachymene eight days after a carbon spot had been placed upon its center. A small part of the spot had been left behind on the flank of the apex toward leaf #2, but the large intact mass had been left on the side of the apex between leaves #1 and #3. The spot had been left on the opposite side of the apex from the presumed position of the next to appear leaf--#5, approximately between #2 and #3. 75x.

of its position (fig. 11), which was opposite the growth center of the future leaf #7 (between leaves #2 and #5). This result could only have occurred by the origin of the foliar primordium under the laterally positioned carbon mass and the subsequent leaving behind of the carbon on the abaxial side of the leaf by its apical growth. Angiosperm foliar primordia usually show apical growth up to a height of several hundred µm (Foster[19]). Probably more frequent were the cases of spreading of the carbon spot while it was being left behind laterally. Fig. 12 shows the spot spread on the sides of the apex, but still above the region of origin of foliar primordia. It had been moved opposite the just appeared primordium #6. Fig. 13 illustrates spread of the spot over one half of the apex in the direction of the youngest visible foliar primordium. It had been moved away from the growth center of the next to appear leaf, #7, between leaves #2 and #4. In fig. 14 the spot had been left behind laterally and had spread circumferentially, ending in the axil of a now advanced foliar primordium, #2. A younger foliar primordium, #5, had arisen radially within the carbon mass. The spot had been moved away from the growth center of the next to appear leaf, #6, between #1 and #3. A similar case is shown in fig. 15, where the carbon spot had become a circumferentially spread mass in the axil of an older primordium (removed) and several younger primordia had arisen radially within it. These lateral or circumferential spreadings of the carbon mass as it is left behind may be interpreted to indicate divisions of underlying cells during the process.

Trachymene coerulea, a member of the Umbelliferae, is advantageous for this kind of work because the shoot apex ends its vegetative development by transforming into a wide, flat inflorescence, and all of its parts may be observed in one plane of focus for photomicrography. The floret primordia arise first on the edge of the flat meristem, and successive ones arise toward the center of the disc. Those apices that had not shown an indication of the transformation, but which were physiologically, if not morphologically determined, often left the intact carbon spot in a lateral position which came to be utilized in the formation of the first floret primordia (fig. 16). In other experiments the carbon had been left behind laterally and spread circumferentially prior to formation of the florets on the edge of the inflorescence axis (fig. 17). If, however, the carbon spot were applied to the center of the apex during the early growth in width leading to the flat disc-like inflorescence apex, little displacement of it occurred, and it remained near the center (fig. 18). Or, if applied after the first floret primordia had appeared on the edge of the disc, the carbon spot remained at the center of the inflorescence apex and was merely spread slightly, either by new cell divisions or by cell expansions (fig. 19).

After the application of two or more carbon spots around the center of the shoot apex, additional information has been obtained

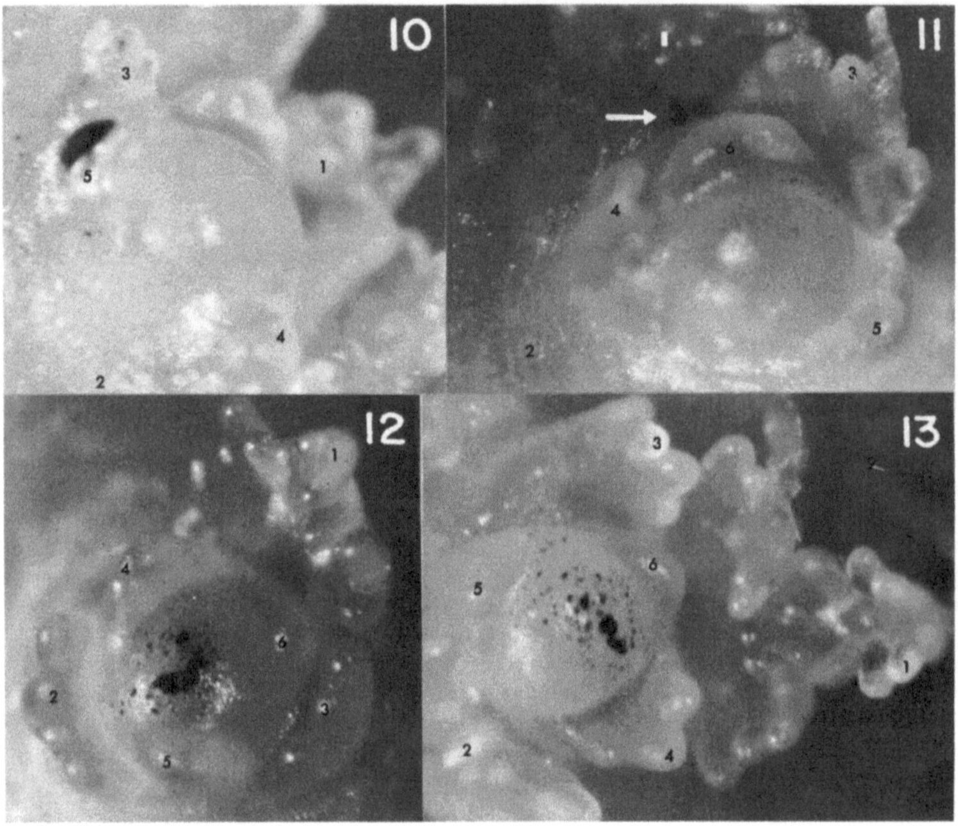

Fig. 10-13.--Fig. 10. Trachymene s.a. 11 da after spot was placed on its center. Spot was left behind laterally where leaf #5 arose --ca. across apex from presumed position of next-to-appear leaf-- #6, between #1 & #4. 75x.--Fig. 11. Trachymene s.a. 8 da after spot was placed on its center. Spot was left behind by rapid growth, and leaf #6 arose under it. By apical growth of latter, the carbon mass was left behind on abaxial side of leaf, and ended in axil of leaf #1 (arrow), as far as possible from position of next-to-appear leaf--#7, between #2 and #5. 75x.--Fig. 12. Trachymene s.a. 8 da after spot was placed at center. Spot is now toward leaf #2 and had spread irregularly, and opposite the new primordium #6, between #1 & #3. 75x.--Fig. 13. Trachymene s.a. 11 da after spot was placed on its center. Spot was left behind, with lateral spreading, over 1/2 of apex, mainly toward youngest visible leaf, #6. It had been moved away from the newest growth center--primordium #7, ca. between leaves #2 and #4. 75x.

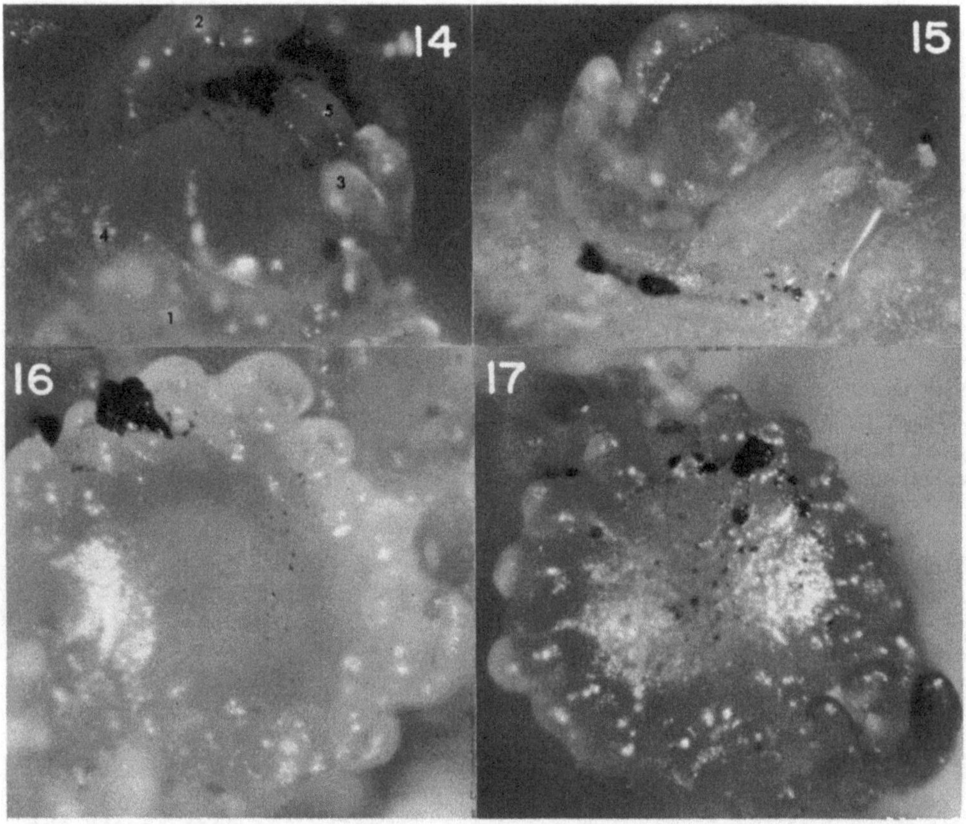

Fig. 14-17.--Fig. 14. Trachymene s.a. 7 da after spot was placed on its center. Spot was left behind laterally and had spread circumferentially, ending mostly in axil of leaf #2. Meanwhile leaf #5 arose within the mass and left behind part of it on its abaxial side, having been moved away from newest growth center, the next-to-appear leaf, #6, between #1 and #3. 75x.--Fig. 15. Trachymene s.a. 11 da after spot was placed on its center. Spot was left behind laterally and spread circumferentially. At least 2 leaves arose within mass and had left it behind on their abaxial surfaces. Spot was finally left in axil of an older leaf (excised). 75x.--Fig. 16. Early inflorescence apex of Trachymene developed from s.a. center-tagged by spot 11 da before. The carbon, now divided circumferentially into 2, was left behind laterally, and the first floret primordia arose beneath it. 75x.--Fig. 17. Early inflorescence apex of Trachymene developed from a s.a. center-tagged by spot 7 da before. During growth that left spot behind laterally, there was much circumferential and irregular growth of meristematic surface, and spot was finally divided into numerous small masses. 75x.

on the sites of cell divisions in the meristem. Growth after application of a carbon spot on both sides of the presumed center of the apex (cf. fig. 3) had caused the leaving behind of both spots in the same lateral position, and leaf #2 had arisen radially between them. Some bits of the innermost carbon spot had spread slightly to the other side of the apex (fig. 20). Three spots placed around the putative center of the apex (cf. fig. 4) were left behind on the first two foliar primordia and in the axil of an older one, respectively (fig. 21). This is a relatively rare case of a functional growth center in the center of the apex; it had produced cells that carried the three spots along different radii toward the edges of the dome. Four carbon spots placed equidistantly around the presumed center of the apex (cf. fig. 5) were all left behind laterally on the same side of the meristem (fig. 22). The same operation performed upon an early stage of the inflorescence showed the spots, with some spreading within each, to have been left behind laterally and almost equidistantly upon four areas of floret primordia (fig. 23). Upon a later stage of the inflorescence, however, the spots originally around the center were mostly left behind at the edges of a not yet floret-producing apex (fig. 24).

Vicia faba.--In this plant (Leguminosae) with distichous phyllotaxy, a single carbon spot placed upon the presumed center of the shoot apex (fig. 2) was always left behind laterally and asymmetrically. The carbon spot may have remained largely intact and was sometimes found on a side of the apex that produced no foliar primordia (fig. 25. Apparently a growth center at the tip of the apex had produced cells that were left behind down this flank. More frequently, such a spot was left in the axil of a recently produced foliar primordium, on a side of the apex morphologically 90° away from the first (fig. 26). Here a growth center that formed the youngest leaf (#2) produced cells that carried the spot across the apex and into the axil of leaf #1. Usually the carbon spot was divided on being left behind by the growth of the shoot apex. Fig. 27 shows the spot divided into three parts, one in the axil of an older leaf, and two on a non-leaf-producing side of the apex. Fig. 28 shows the spot divided into four parts and left behind in the axil of the next to youngest foliar primordium. While the leaving behind of the carbon on the flank near leaf #2 was caused by cells produced by the growth center of leaf #3, the breaking up of the spot was due to cell divisions in various directions. Fig. 29 shows the spot divided into numerous small parts and left behind largely in the axil of the older foliar primordium. Fig. 30 shows the spot divided into innumerable small parts and spread circumferentially in the axil of the older foliar primordium. Probably the growth center that had formed leaf #1 produced relatively few cells toward the center, and thus the spot was carried little from its original position. Breaking up of the spot on the side next to leaf #2 was probably due to cells being produced toward the non-leaf producing flanks, or by the early stages of a growth center for the next-to-

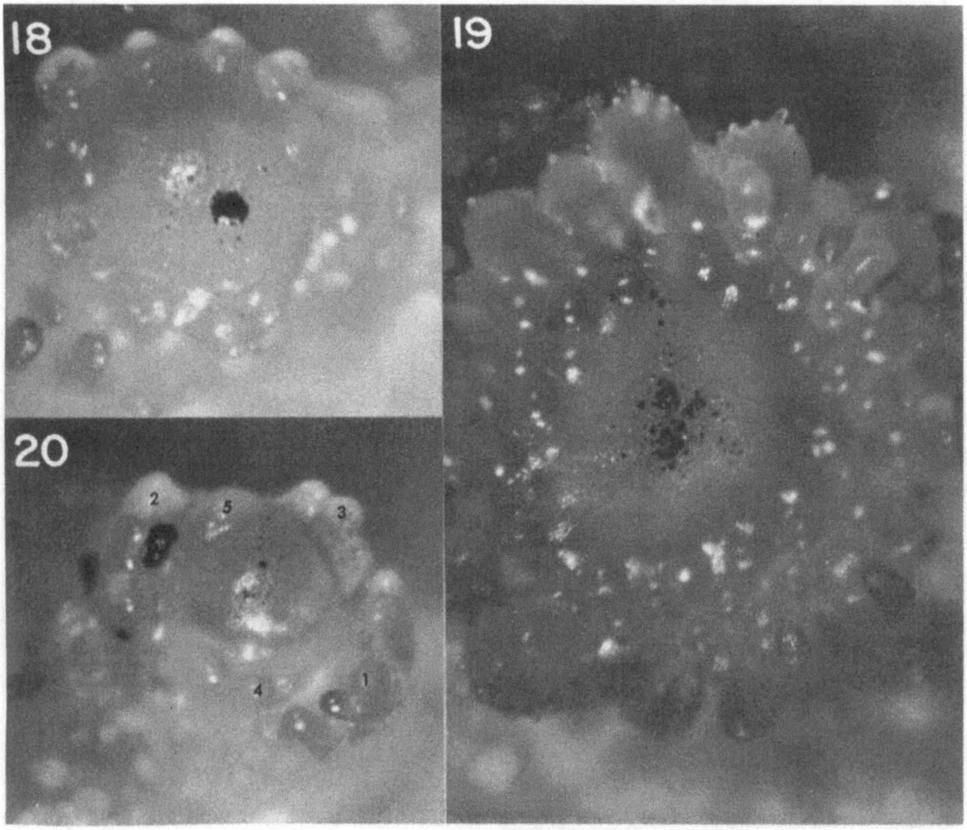

Fig. 18-20.--Fig. 18. Inflorescence apex of Trachymene that developed from s.a. upon center of which spot was placed 8 da before. At time of application of spot, s.a. had already shown beginnings of disc-like spreading, and was thus in early transformation to an inflorescence. Showing that the inflorescence disc mostly grows by uniform cell formation, the spot was left behind only slightly from center. 75x.--Fig. 19. Inflorescence apex of Trachymene that developed from a later stage already with the first marginal florets; the carbon spot that was then placed at its center 7 da before had been merely spread slightly in all directions either by divisions or enlargement of cells. 75x.--Fig. 20. Trachymene s.a. 9 da after 2 spots were placed around its presumed center. A growth center for future leaf #6 (between #1 and #4) had developed lateral to both spots and they were left behind together upon opposite side of the apex, where leaf #2 had arisen radially between them. 75x.

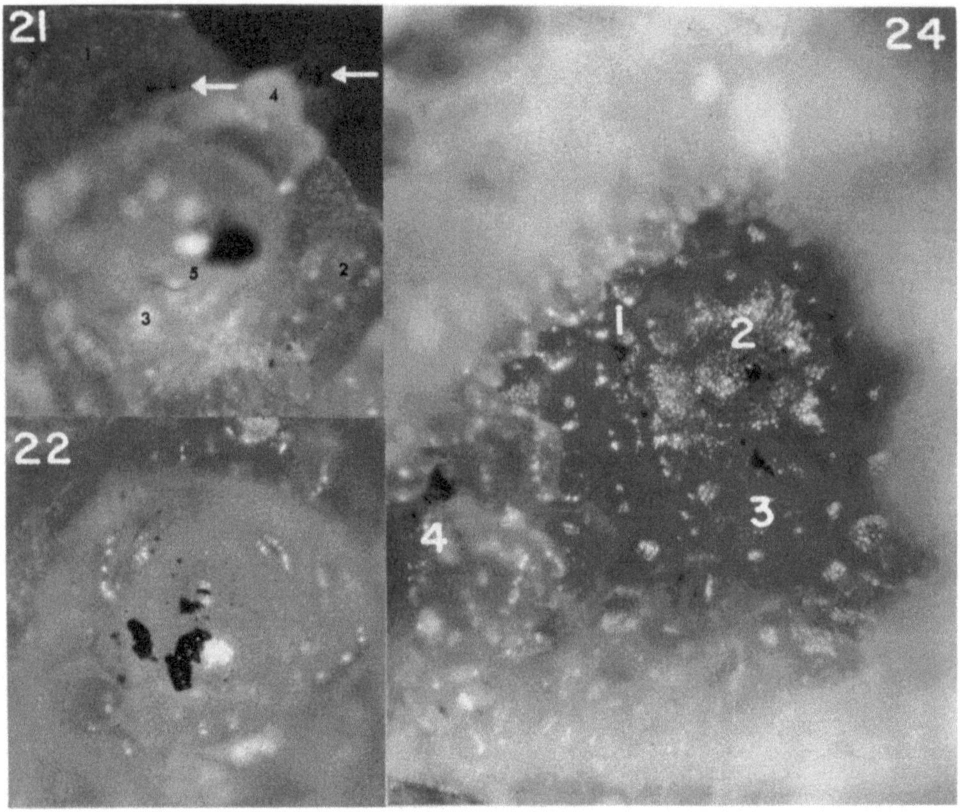

Fig. 21-24.--Fig. 21. Trachymene s.a. 9 da after 3 spots were placed around its center. Apex actually had a center of symmetrical cell formation at center, and the spots had been left behind laterally along opposed radii. One ended in axil of leaf #1 (excised) and was slightly spread circumferentially (white arrow). Another ended atop leaf #4 where it had been spread into several small bits (white arrow), and the third was partly atop leaf #5. 75x.--Fig. 22. Trachymene s.a. 9 da after 4 spots were placed around its center. A growth center had functioned lateral to all 4 spots and they had been left behind, with some changes in their relation to each other, on one side of the apex. 75x.--Fig. 24. Trachymene inflorescence apex that, at time of application of 4 spots around its center, had already produced the first marginal florets, 9 da after operation. Only one of the spots (#4) had been left behind in area of new floret production; the other 3 are only at the edges of the not yet-floret-forming apex. Obviously a growth center, within the 4 spots, had produced cells along all radii, carrying the spots apart. 75x.

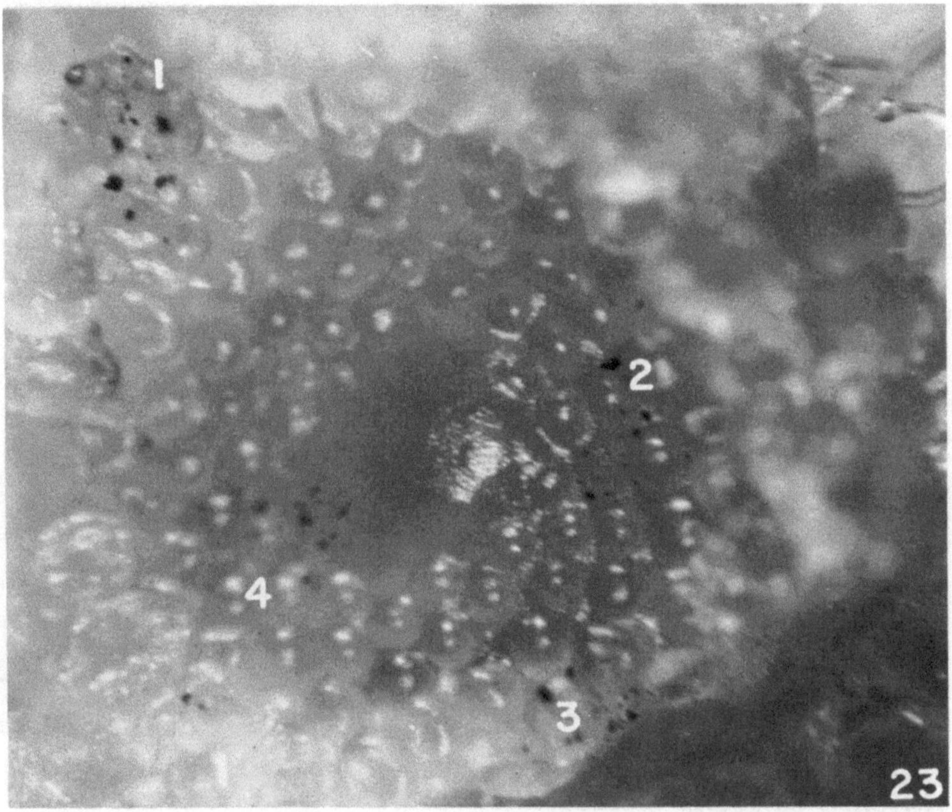

Fig. 23. Inflorescence apex of Trachymene nine days after four carbon spots had been placed around its center. At the time of application, no floret primordia had been formed. Spots (1), (2) and (3) had been left behind entirely in the lateral floret-forming regions, while spot (4) was only partly there, since it had been divided into an inner part that was upon the not yet floret-forming region of the apex. A growth center within the four spots had produced cells almost equally along all radii. 75x.

appear leaf.

Two spots applied parallel to the two rows of leaves (cf. fig. 6) were both sometimes left behind together in the axil of the younger foliar primordium (fig. 31). Though the newest leaf had not yet appeared, its growth center, next to leaf #1, had obviously functioned rapidly to produce cells that carried both spots across the apex and into the axil of leaf #2. When the spots were applied around the center and above the non-leaf-producing sides of the apex (cf. fig. 7), they were slowly left behind on the flanks (fig. 32) below the level of origin. This slowness contrasts with the rapidity of displacement of the spot in the line of the foliar primordia (cf. fig. 31).

DISCUSSION.--The pattern of cell formation in the shoot apex shows some variability. In keeping with the anatomical importance of the leaf in shoot development (cf. Esau[18]) here it is evident that the growth centers in the meristem leading to the origin of successive leaves are the main sources of new cells. The positions of these growth centers coincide with the phyllotaxy. In _Trachymene_, which has a helical leaf arrangement, the newest growth center is at a prolongation of the helix. In _Vicia_, with its distichous leaf arrangement, the newest growth center is 180° across the shoot apex from the youngest visible leaf. A valid interpretation of the growth pattern is that an overall organismal control exists which determines the sites and directions of cell division. This conclusion agrees with Vöchting's[48] thesis that the fate of a cell is a function of its position. Thus, a carbon spot placed on the presumed center of the shoot apex was always left behind in a lateral position, obviously carried there by cells that had been produced by an adjacent growth center, and at the end of the observations was on the flank of the apex, in the axil of a foliar primordium, on the surface of a foliar primordium, or, finally, beyond the innermost foliar primordia, upon, or associated with an older leaf that was a considerable distance below the dome of meristem. The placing of two or more carbon spots around the presumed center of the shoot apex was done with the view of ascertaining whether the central cells produced new cells that would cause the carbon spots to be left behind symmetrically upon the apex. Such a result sometimes did occur; the spots were leaf behind opposite to each other and were found down the flanks of the apex in comparable positions to those resulting from the application of a single spot to the apex. Obviously, new cells had been produced symmetrically between the spots. Considering the fact that most cell production of the angiosperm shoot apex concerns leaf origin and development, it may be suggested that the rare cases of a growth center in the center of the apex constitutes origin of stem tissues. However, in most cases all the spots were left behind on the same side of the apex; this could only have happened when new cells had been produced by a growth center lateral to all the spots . If these growth centers

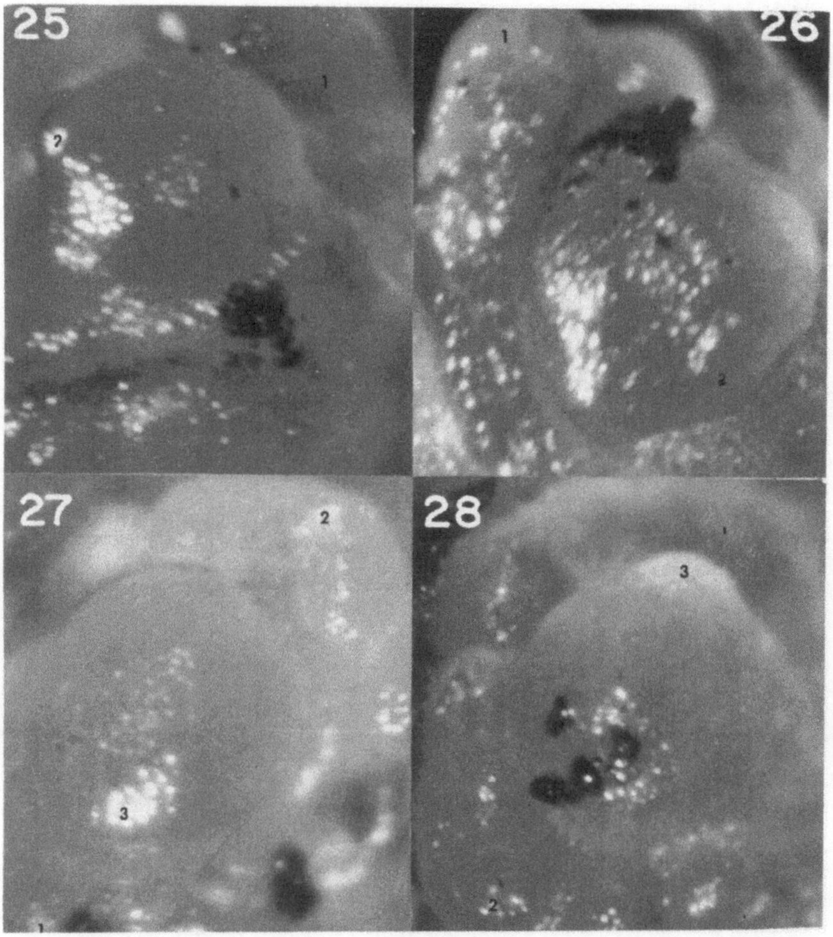

Fig. 25-28.--Fig. 25. Shoot apex of Vicia 20 da after one carbon spot was placed at center; spot remained largely intact and was left behind on flank below non-leaf-forming side. #2 is youngest leaf and #1 the next. 150x.--Fig. 26. Shoot apex of Vicia 9 da after 1 carbon spot was placed at center. Spot remained largely intact and was left behind in axil of older leaf, #1. Obviously, a growth center functioning in formation of leaf #2 had caused apex to grow in manner that would leave behind the upper edge of the apex (including "center" where spot had been placed). 150x.--Fig. 27. Vicia s.a. 15 da after spot placed at its center. Spot was left behind laterally, and divided; 2 masses were at base of flank of a non-leaf-producing side of apex, and 1 was in axil of leaf #1. Apparently a growth center near spot had caused these displacements before appearance of younger leaf #3. 150x.--Fig. 28. Vicia s.a. 13 da after spot was placed at center; spot was divided into 4 parts and left behind toward next-to-older leaf, #2. This direction of growth was obviously result of development of growth center of younger leaf, #3. 150x.

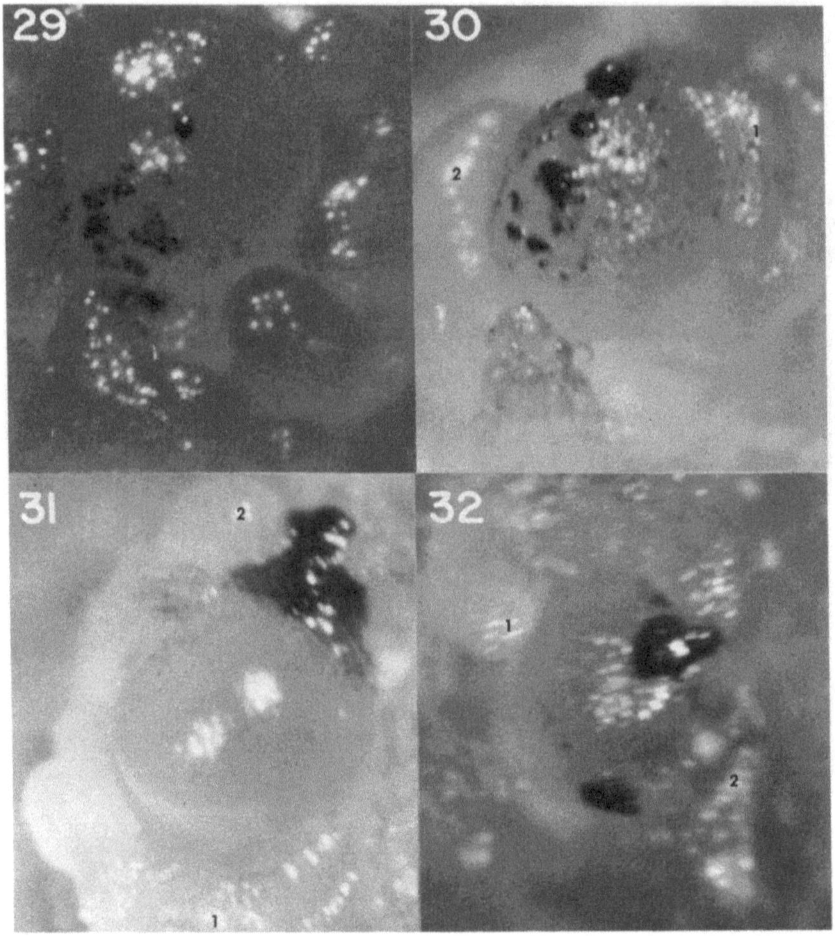

Fig. 29-32.--Fig. 29. Vicia s.a. 7 da after 1 carbon spot had been placed at center; spot was left behind laterally and divided into many small masses and ended in axil of an old primordium. No young primordium was yet visible on opposite side of s.a., where obviously a growth center had functioned. 150x.--Fig. 30. Vicia s.a. 7 da after putting 1 spot on its center; spot was left behind with much spreading, mostly in axil of the older leaf, #2, while essentially none of it was in axil of younger, #1. Probably the last stages of growth center producing #1 had left carbon behind on side of apex near #2, where it was then spread by first stages of growth of next-to-appear leaf. 150x.--Fig. 31. Vicia s.a. 7 da after 2 spots had been applied around center parallel to line of leaves (cf. Fig. 6). A growth center had developed near leaf #1 which left both spots behind in axil of opposite leaf, #2.--Fig. 32. Vicia s.a. 7 da after putting 2 carbon spots around its center at rt. angles to 2 rows of leaves (cf. Fig. 7). These spots, above the 2 non-leaf-producing sides of apex had been left behind only half-way down the sides of the apex. 150x.

are considered sites of "initial cells" then Foster's[21] suggestion that there are probably no definite, unchangeable "initial cells" in the shoot apex is supported. In the present work, the carbon spots placed on the shoot apices were left behind laterally within seven to fifteen days, and at the end of such periods were often upon or among foliar primordia that had been produced on the flanks. In contrast, Loiseau[31] found that a centrally placed spot was left behind three-fourths of the way down the flank in fifty-five days. Perhaps this comparative slowness was due to growth inhibitions from the "media" (albumin glue, butter, peanut oil) used to suspend the lampblack. He thought that spots were left behind slowly from the center, faster from the flanks of the apex, and considered that this correlated with his previous finding of rates of mitosis in the two regions. However, in abnormally small apices (young plants?) a centrally placed spot was rapidly left behind onto the flanks. While spots were being left behind, they showed both radial and tangential (circumferential) deformations.

In the present work the spreading of the carbon spots during their being left behind on the apex may be regarded as results of supplementary divisions that are not directly related to the growth centers.

The usual occurrence of the growth center upon the side of the meristem is not to be construed as support for Buvat's[9] thesis that the lateral tissues of the shoot apex are the sole source of new cells in the meristem. He termed this region the "meristem ring" and found it to consist of small, rapidly dividing cells. These cytological and histological features of the lateral tissues of the shoot apex have been confirmed in paraffin sections by many investigators (cf. Gifford and Corson[22]), yet there is a body of evidence, up to now neglected, that illustrates the fallacy of considering the lateral region as the main cell-producing region of the shoot apex. This is the results of ionizing-radiation bombardements of the shoot apex (cf. Lapins and Hough[30] and the pertinent literature cited). The results are uniform; after sublethal irradiation, the central cells (putative quiescent center, according to Buvat[9]) are destroyed and the lateral cells (putative meristem ring of Buvat[9]) survive and may regenerate one or more functional shoot apices. The present author submits that this result is the opposite to that to be expected if the central cells are truly quiescent and the lateral cells really rapidly dividing. According to the guiding principle of radiotherapy established long ago by Bergonié et Tribondeau[7], the rapidly dividing cells (lateral, or flank tissues, the "anneau méristèmatique of Buvat[9]) should have been destroyed, and the central cells (méristème d'attente" of Buvat[9]) should have been resistant.

In the root apex, where Clowes[11-15] has shown that a true quiescent center exists, sublethal irradiation treatments yield results in agreement with the principle of Bergonié et Tribondeau[7]; the lateral, rapidly dividing cells are destroyed, but cells of the quiescent center are resistant and later divide to reestablish a functional root apex.

An alternative explanation of these conflicting results may be had in the work of Sparrow and Woodwell[46]. They suggest that the radiosensitivity of the plant cell to ionizing radiation varies directly with the volume of its nucleus, and inversely with the rate of nuclear division. It is interesting that their ideas constitute a contradiction to the principles of Bergonié et Tribondeau[7], and of the well established results of Clowes[11-15] on the root apex. However, it is conceivable that the shoot apex, a part of the aerial plant system, differs in this fundamental aspect from the root. If so, then the large-nucleated central cells of many angiospermous shoot apices would have the requisites of high radiosensitivity, while the lateral tissues of the same meristem would show radioresistance. Furthermore, if the central cells with large nuclei are, in fact, highly radiosensitive, the utilization of radioisotope containing compounds, e.g., thymidine-H^3, could affect them by stopping all their divisions, as demonstrated by Steeves et al[47], since tritium has a soft, short range β radiation that would be very efficiently absorbed by the tissues. Such a suggestion casts doubt upon the validity of the autoradiographic technique as currently applied to shoot apices (cf. Lance-Nougarède et Bronchard[29] and Bernier and Jensen[8]). Other results of Steeves et al[47] utilizing microspectrophotometry showed that the putative central zone of the apex produced DNA, indicating probable mitoses in its cells.

A valid interpretation of the present results is that the center of the shoot apex sometimes is a main source of new cells, but it can be displaced by laterally oriented new centers of growth. This tentative conclusion is in conflict with much recent work (cf. Gifford and Corson[22]) that apparently supports Buvat's[9] thesis that the upper central portion of the shoot apex is to be regarded as a quiescent center where divisions of cells are infrequent. This recent work consists of counts of mitoses in paraffin sections, measurements of DNA origin and occurrence by autoradiography of paraffin sections and various histochemical procedures mostly on paraffin sections. Buvat's[9] thesis, and the subsequent work that supports it, suffers from the negating shortcoming that it shows only the state of a section of the shoot apex at a particular instant after it had been modified by the necessary treatments involved in exposure to a radioisotope, excision, killing, dehydrating, sectioning, and staining. No sequential observations are possible upon the same apex by these standard cytological, autoradiographic, or histochemical techniques; thus, what occurs in one apex during a specified interval of time cannot possibly be ascertained by their use. These techniques make use of the only alternative-- the seeking by the investigator of what he decides subjectively to be sequential stages within possibly scores of serially sectioned shoot apices involving thousands of sections.

It may be provisionally concluded that, by virtue of the fact that sometimes the center of the shoot apex happens to be a growth center, during the variable location of growth centers in the meristem, the central cells divide and produce new cells.

If there is a quiescent center in the apex of Trachymene coerulea or of Vicia faba (as is described for the latter by Lance[28]) the present findings indicate that it must be constantly moving out of the tip and constantly being replaced. The possibility must be considered that the central region, with its special cytological characteristics, as described from paraffin sections, is a fixation artefact.

Thus, the classical concept of "central initials" has no meaningk according to the present data, since any central cells of these apices are only temporarily located there. Nevertheless, the problem of the initial cell or cells in the shoot apex must be reckoned with. Beard[5] used X-irradiation and genetic analyses to ascertain that most mutant branches in flax arose from a single initial cell. The results of Howard[25] in which evidence is accrued for the numbers of initial cells in the layers of the shoot apex of potato is an example of genetic research that indicates the desirability of orientation analysis of the terminal meristem. If L_1 produces only epidermis, and L_2 is the site of origin of foliar primordia (Howard[25]) it is important to know how these layers are maintained if their growth centers constantly shift about, as the present evidence indicates. Where are "initials" located in such a constantly changing cell population? Is it perhaps logical to suggest that any cell of a particular layer in the shoot apex may function at any time as an "initial" if it happens to be encompassed in a new growth center? The recent work of Chen and Ross[10] and Ross and Atkinson[40] indicates that the above suggestion may be valid. They studied shoot apices of tetraploid sorghum that had been treated with colchicine. They described probable somatic meioses and showed that a new diploid shoot apex could arise from one or more of the 2n cells. The diploid cells, they found, had a competitive advantage over the tetraploid, and quickly replaced the latter and produced a functional 2n shoot apex. These "initial" diploid cells could arise in any part of the original 4n apex, tip or flank, and function to produce a new apex. Since the diploid cells were obviously new growth centers, they may be compared with those described in the present work after tagging experiments. An analogous situation was described by Ball[2] after excision or destruction of the top central cells of the shoot apex of Lupinus albus. The remaining lateral cells of the dome of meristem, though they had previously functioned as sites of origin of new leaves, became new growth centers and regnerated normal functional shoot apices.

The transition of the shoot apex of Trachymene coerulea into an inflorescence apex has made it possible to test another tenet of

Buvat's[9] thesis: that the quiescent center of the shoot apex is being held in reserve for future transformation into a floral apex. The present results show that the tagged central cells of an early stage of the transformation were left behind laterally and were utilized in the formation of peripheral floret primordia on the disc-shaped inflorescence apex. Thus, lateral regions of the wide, flat inflorescence disc thus owe their origin, in part at least, to cells that were formerly at the center of the transforming shoot apex. The displacement of these former central cells toward the periphery of the meristem by new cells produced by a lateral growth center indicates that the development of the early inflorescence apex is basically similar to that of the shoot apex. Thus, there is never any passive center of the shoot apex at the time of transformation. Only when the central region of a later stage of inflorescence development was tagged, did it remain approximately in its original position, but even then it expanded, separating the carbon spot into numerous small particles. It may be concluded that there is no quiescent center in these shoot apices, and furthermore, since the cells of the central region are being constantly changed, there is no conserving of them for future transformation into an inflorescence.

REFERENCES

1. Ball, E. (1944) The effects of synthetic growth substances on the shoot apex of Tropaeolum majus L. Amer. J. Bot. 31:316-327.

2. Ball, E. (1950) Regeneration of the shoot apex of Lupinus albus after operations upon the central initials. Science 112: 16-17.

3. Ball, E. (1962) Studies of living shoot apices. IN: Plant Tissue Culture and Morphogenesis, A Symposium of the American Society of Plant Physiologists, Jacksonville, Fla. pp. 47-77. New York.

4. Bannan, M. W. (1956) Some aspects of the elongation of fusiform cambial cells in Thuja occidentalis L. Can. J. Bot. 34:175-196.

5. Beard, B. H. (1970) Estimating the number of meristem initials after seed irradiation: a method, applied to flax stems. Rad. Bot. 10: 47-57.

6. Bergann, F. und L. Bergann. (1962) Über Umschichtungen (Translokationen) an den Sprossscheiteln periklinaler Chimären. Der Züchter 32: 110-119.

7. Bergonié, J et L. Tribondeau. (1906) Interprétation de quelques resultats de la radiothérapie et essai de fixation d'une technique rationelle. C. R. Acad. Sci. Paris 143: 983.

8. Bernier, G. and W. A. Jensen. (1966) Pattern of DNA synthesis in the meristematic cells of Sinapis. Histochemie 6: 85-92.

9. Buvat, R. (1952) Structure, évolution et fonctionnement du méristème apical de quelques Dicotylédones. Ann. Sci. Nat. Bot. 11e Sér. 13: 199-300.

10. Chen, C. H. and J. G. Ross. (1965) Colchicine-induced somatic chromsome reduction in sorghum. V. Diploidization of the stem apex after treatment of tetraploid seedlings. Can. J. Genet. Cytol. 7: 21-30.

11. Clowes, F. A. L. (1956) Localization of nucleic acid synthesis in root meristems. J. Exp. Bot. 7: 307-312.

12. Clowes, F. A. L. (1958) Development of quiescent centres in root meristems. New Phytol. 57: 85-88.

13. Clowes, F. A. L. (1961) Effects of β-radiation on meristems. Exp. Cell Res. 25: 529-534.

14. Clowes, F. A. L. (1963) X-irradiation of root meristems. Ann. Bot. NS 27: 343-352.

15. Clowes, F. A. L. (1965) Meristems and the effect of radion on cells. Endeavour 24: 8-12.

16. Dingler, H. (1886) Zum Scheitelwachsthum der Gymnospermen. Ber. deut. bot. Ges. 4: 18-36.

17. Douliot, H. (1890) Recherches sur la croissance terminale de la tige des Phanérogames. Ann. Sci. Nat. Bot. 7e Sér. 11: 283-350.

18. Esau, K. (1965) Plant Anatomy. Wiley, New York.

19. Foster, A. S. (1936) Leaf differentiation in Angiosperms. Bot. Rev. 2: 340-372.

20. Foster, A. S. (1939) Problems of structure, growth and evolution in the shoot apex of seed plants. Bot. Rev. 5: 454-470.

21. Foster, A. S. (1941) Comparative studies on the structure of the shoot apex in seed plants. Bull. Torr. Bot. Cl. 68: 339-350.

22. Gifford, E. M. Jr. and G. E. Corson Jr. (1971) The shoot apex in seed plants. Bot. Rev. 37: 143-230.

23. Hanstein, J. (1868) Die Scheitelzellgruppe im Vegetationspunkt der Phanerogamen. Festschr. Niederrhein Ges. Natur. -u. Heilkunde: 109-143.

24. Hofmeister, W. (1857) Beiträge zur Kenntniss der Gefässkryptogamen. Abh. Kön. Säch. Ges. Wiss. 3: 603-682.

25. Howard, H. W. (1967) Experiments on X-ray irradiation of potatoes: analysis of chimeras and differentiation effects. IN: Induzierte Mutationen und ihre Nutzung. Erwin-Baur-Gedächtnisvorlesungen IV. 1966, pp. 311-317. Akademie-Verlag, Berlin.

26. Koch, L. (1891) Ueber Bau und Wachstum der Sprossspitze der Phanerogamen. I. Die Gymnospermen. Jahrb. wiss. Bot. 22:491-680.

27. Korschelt, P. (1884) Zur Frage über das Scheitelwachstum bei den Phanerogamen. Jahrb. wiss. Bot. 15: 642-674.

28. Lance, A. (1953) Sur l'absence d'initiales apicales et la configuration de l'anneau initial chez Vicia faba L. C. R. Acad. Sci. Paris 236 (part 1): 510-512.

29. Lance-Nougarède, A. et R. Bronchard. (1965) Métabolisme des acides nucléiques dans le méristème apical du Perilla nankinensis au cours des diverses phases du developpement. C. R. Acad. Sci. (Gr. 11) 260: 3140-3143.

30. Lapins, K. O. and L. F. Hough (1970) Effects of gamma rays on apple and peach leaf buds at different stages of development--II. Injury to apical and axillary meristems and regeneration of shoot apices. Rad. Bot. 10: 59-68.

31. Loiseau, J.-E. (1962) Activité mitotique des cellules superficielles du sommet végétatif caulinaire. Mém. Soc. Bot. Fr. pp. 14-23.

32. Lyndon, R. F. (1968) Changes in volume and cell number in the different regions of the shoot apex of Pisum during a single plastochron. Ann. Bot. 32: 371-390.

33. Lyndon, R. F.(1970a) Rates of cell division in the shoot apical meristem of Pisum. Ann. Bot. 34: 1-17.

34. Lyndon, R. F. (1970b) Planes of cell division and growth in the shoot apex of Pisum. Ann. Bot. 34: 19-28.

35. Lyndon, R. F. (1970c) DNA, RNA, and protein in the pea shoot apex in relation to leaf initiation. J. Exp.Bot. 21: 286-291.

36. Lyndon, R. F. (1971) Growth of the surface and inner parts of the pea shoot apical meristem during leaf initiation. Ann. Bot. 35: 263-270.

37. Nägeli, C. (1878) Ueber das Scheitelwachsthum der Phanerogamen. Bot. Zeit. 36: 124-126.

38. Plantefol, L. (1947) Hélices foliaires, point végétatif et stèle chez les dicotylédones. Rev. Gén. Bot. 54: 1-32.

39. Pringsheim, N. (1869) Über die Bildungsvorgänge am Vegetationskegel von Utricularia vulgaris. Monatsber. Kön. Preuss. Akad. Wiss. 92-115.

40. Ross, J. G. and G. F. Atkinson. (1965) Colchicine-induced somatic chromosome reduction in sorghum. VI. Nuclear volume changes within diploid stem apices after colchicine application. Can. J. Genet. Cytol. 7: 31-36.

41. Satina, S. and A. F. Blakeslee. (1941) Periclinal chimeras in Datura stramonium in relation to development of leaf and flower. Amer. J. Bot. 28: 852-871.

42. Schmidt, A. (1924) Histologische Studien an phanerogamen Vegetationspunkten. Bot. Arch. 8: 345-404.

43. Schoch-Bodmer, H. und P. Huber. (1945) Das Spitzenwachstum der Fasern bei Linum perenne L. Experientia 1: 327-328.

44. Schüepp, O. (1926) Meristeme. Linsbauer's Handbuch der Pflanzenanatomie, IV.

45. Soma, K. and E. Ball. (1963) Studies on the surface growth of the shoot apex of Lupinus albus. IN: Meristems and Differentiation, Brookhaven Symposia in Biology No. 16, pp 13-45.

46. Sparrow, A. H. and G. M. Woodwell. (1962) Prediction of the sensitivity of plants to chronic gamma irradiation. Rad. Bot. 2: 9-26.

47. Steeves, T. A., M. A. Hicks, J. M. Naylor, Y. P. Rennie. (1969) Analytical studies on the shoot apex of Heliantus annuus. Can. J. Bot. 47: 1367-1375.

48. Vöchting, H. (1878) Über Organbildung im Pflanzenreich. Max Cohen, Bonn. 258 pp.

-DISCUSSION-

D'AMATO - Are you sure that the lanolin, which is lipid soluble, does not move spontaneously? This is the impression I have from some of your photographs. Have you done experiments on non-dividing tissues?

BALL - We have not performed such experiments with lanolin, but we have used other compounds such as the very inert Dow-Corning stopcock grease and get similar results. I should point out that spotting with these compounds does not interfere in any way with the phyllotactic pattern of the meristem or of its subsequent development. I don't think these compounds spread, but we plan to do some time-lapse photography to check this point.

HABER - But in some cases where you put more than one spot on they eventually moved closer together which seems to me to indicate surface spreading.

BALL - No, one can get cell division on one side of the spot group accompanied by growth beneath the spots which may give this impression. I really don't think spreading of the sort you are suggesting really occurs.

VAN'T HOF - Your earlier experiments, where you puncture a single cell, showed conclusively that the meristem does move about and this is independent of possible lanolin spreading.

BALL - Yes, it must be emphasized that the movement of a killed cell on the shoot apex is exactly like that of a spot of lanolin.

DAVIDSON - If you put lanolin paste on a *Vicia* leaf in a greenhouse, it certainly doesn't spread even if the temperature is very warm. So, I doubt if it's likely to be spreading here.

CUTTER - Some of your pictures suggest that the growth center of the leaf primordium is much more extensive than the leaf primordium itself.

BALL - Yes, and I intend to investigate this further.

TORREY - I disagree with your diagrammatic representation of the meristeme d'attente (quiescent center). I don't think it reaches the surface at all, but is embedded in the apical region.

BALL - I'm sure they include the surface layers. (See Buvat, R. Ann. Sci. Nat. Bot. 11^{e} ser., 13:199-300, p 228 and 251.)

CONGER, A. - But, the divisional regions are below the surface, and I am not sure that one can deduce the behavior of this underneath

CONGER, A. - region from what is happening on the surface.

BALL - The divisional areas include the surface layers and these cells do not slip over those below.

CONGER, A. - Imagine that I put my finger inside a balloon, and my finger is growing and I put a spot on the surface of the balloon near the apex. Now, imagine that there is only differential elongation on the surface of the balloon (which is the only thing you can detect). It seems to me I can move the spot any way I choose, regardless of whether my finger is moving or staying still, and the spot movement will not provide information on the finger movement.

BALL - I don't think the balloon is a very good analogy of the shoot apex.

SAVAGE -- The whole thing depends on whether this growth center extends to the surface. If you were to invert a root, having removed the root cap and put a spot near the apex, it would move and spread out, but this would not really tell you anything about the activities within the meristem and quiescent center.

JUNIPER - Could I make a point here? There exists a complete cuticle over even the smallest apex, and what you may be following is the synthesis of cutin and not the patterns of cell division.

BALL - I don't think the cuticle could move over the cells which produce it, if I understand cutin production. Hall's (Science, 158:505, 1967) findings on wax exudation indicate the stability of position of cuticle (with its microchannels through which wax is forced out) with regard to the outer wall of the cell. Bolliger (J. Ult. Res., 3:105, 1959) showed that cuticle is made up of lipid particles that have migrated through the outer wall and deposited firmly in the irregularities of that surface. Gifford and Stewart (J. Cell Biol., 33:131, 1967) show the cuticle on the shoot apex to be a closely adhering dark layer (after $KMnO_4$ or OsO_4-glutaraldehyde fixations) over the meristem's outer wall. Finally, the severity of the techniques of isolation (i.e., removal) of cuticle from plant surfaces indicate the great force of adhesion between the two (cf. Kisser, Cytologia, 2:283, 1931; Orgell, Pl. Phys., 30:78, 1955).

CLOWES - Your cinefilms confirm that surface cells actually move; but they can't move independently of the cells below because cells don't slip over each other.

BALL - Yes, this is a tenet of plant growth.

D'AMATO - In actual fact, the French school do not say that the cells of the meristeme d'attente do not divide, but that they rarely divide, and this is important. The cells only stain lightly

D'AMATO - with Feulgen, and also for histones; and I think that this part of the apex may be important in macromolecular synthesis to provide the material for the establishment and maintenance of the apical structure.

ON DETERMINATION OF LEAF PRIMORDIA IN OSMUNDA CINNAMOMEA L.*

CHARLES CARROLL KUEHNERT

Department of Biology, Syracuse University,

Syracuse, New York 13210, USA

INTRODUCTION

Leaf primordia in the vascular plants are formed at regular intervals at the flanks of the shoot apex. The primordia arise as the result of divisions in groups of cells of the surface and subsurface cell layers at the flanks of the apex. In the ferns, the primordium develops a leaf apical cell characteristic of the species, assumes a dorsiventral symmetry, and by rapid growth quickly overtops the apical meristem.

A unique characteristic of the cinnamon fern, Osmunda cinnamomea L., is that with the establishment of the primordium as a distinct morphological entity at the shoot apex, there is no concomitant establishment of an axillary bud which is so characteristic of the flowering plants. Under natural conditions, therefore, during the protracted period of time required for leaf maturation (four years on the average for mature plants), no bud-forming potential is ever found to be associated with the developing leaf.

However, Steeves[20,21,22] showed in Osmunda cinnamomea that of the ten youngest primordia, P_1-P_{10}, the youngest, P_1, when isolated and cultured always developed as a shoot (bud), whereas isolated, cultured P_{10}'s always developed as leaves. He noted that primordia 2 through 9 of this sequence showed a variable response when isolated, with progressively older primordia giving an increasing proportion of leaves. Apices of primordia 2 through 9 which generated shoots appeared to divide into two equal meristematic mounds. One of these developed into a shoot apical meristem according to Steeves[22], whereas the other continued to develop as a leaf. From these investigations, Steeves concluded that primordia produced at the shoot apex of the cinnamon fern are not irreversibly determined as leaves from their inception; that, in point of fact, they remain undetermined for a relatively long period of time. Kuehnert and Steeves[18] reached a similar conclusion. They reported cultured saggital fragments of primordia 4 through 9 showed the capacity to develop as whole shoots or whole leaves.

As regards the interpretation of the concept of determination, a difference of opinion has recently come to light. Cutter[7] argues that in the fern Dryopteris aristata Druce determination of a primordium as a leaf coincides with the establishment of a distinct apical cell (Cutter[6]); and she states that this occurs by the time an undetermined primordium has become a P_3. She further states that if, before determination as a leaf occurs, the primordium is surgically isolated by the proper microsurgical incisions, the whole of the primordial apex will become transformed into a shoot apex and a bud will develop where a leaf was expected.

Cutter's main contention with the interpretation of Steeves, and Kuehnert and Steeves of the phenomenon of leaf determination is whether, in Osmunda, the whole of the leaf apical meristem participates in conversion of the primordium into a shoot. She contends that "...the buds which developed must have been formed from meristematic tissue comprised in the leaf primordium but not from its apex, or, at least, not from the whole of it". She goes on to state, "The criterion for leaf determination must surely be whether the apex of the isolated primordium develops as a leaf or not" (Cutter[7]).

The points of disagreement concerning determination of a primordium as a leaf primordium, in O. cinnamomea, led us to a re-investigation of the problem. One of our aims was to determine the exact nature of the shoots produced by certain of the youngest primordia (P_1-P_4) of the cinnamon fern. A second aim was a better definition of the developmental stage at which determination of a primordium, produced at the shoot apex, is accomplished. That is, at what developmental stage is the primordium irrevocably determined as a leaf?

MATERIALS AND METHODS

Dormant rhizomes of Osmunda cinnamomea L. were held in a cold room at 1-3°C. As needed, the rhizomes were trimmed of excess root material and dead leaves until they consisted of a short piece of stem (ca. 0.5 cm) and the apical bud (Haight and Kuehnert[8]). The apical bud contained the shoot apical meristem

(SAM) and the remaining annual leaf sets (Steeves and Wetmore[26]). The buds were surface sterilized in a 7% (w/v) solution of Percloron (70% calcium hypochlorite; Pennsalt Chemicals Corporation, Philadelphia, Pa.) and rinsed in two changes of sterile distilled water.

The buds were removed to a sterile transfer room where, with the aid of a stereodissecting microscope, the remaining primordia were prepared for culture. The primordia cultured were isolated from the remainder of the shoot system in one of two ways: (1) All remaining primordia except a retained P_1, P_3 or P_4 were removed from the bud. A block of meristematic tissue was excised which consisted of the shoot apical meristem (SAM) plus the desired retained primordium. Final dimensions of the meristematic plug were approximately one cubic centimeter. The plugs were held on sterile agar plates at 4°C for 24-48 hrs. After cold incubation, those plugs which showed surgical damage to either the SAM or any of the retained primordia were discarded. Primordia on the remaining plugs were then incised with a microscalpel, the incision being made between the shoot apical meristem and the retained primordium. The incision extended through the incipient vascular tissues and into the immature pith tissue of the plug. In certain instances, the primordium was further isolated by a piece of sterilized mica which was inserted into the incision. During these microsurgical manipulations no attempt was made to destroy incipient primordia. (2) Certain primordia in the series were simply excised from the bud, and cultured as separate units, or as pairs in physiological contact.

All units were cultured in vessels containing approximately 15 ml of a modified Knudson's medium (Steeves, Sussex and Partanen[25]) plus 2% (w/v) sucrose and 0.8% (w/v) agar. The culture units were maintained in controlled environment growth chambers for 3, or 8 weeks at a temperature of 24° ± 1°C and an illumination of 75-150 ft-c supplied by a mixture of cool white fluorescent tubes and incandescent bulbs. The photoperiod was maintained as an alternating 12 hr light:12 hr dark period.

A stereodissecting microscope was used in evaluating morphologic developmental characteristics of the cultured units. When pertinent, these units subsequently were prepared for histological examination by the usual histological methods, (Jensen[11]; Johansen[12]).

All experiments were repeated at least twice. Experimental results have been shown to be consistently repeatable through several replications. Data reported represent pooled results of at least three separate experiments.

RESULTS

From these investigations it has been clearly shown (Haight and Kuehnert[8]) that: (1) adaxial, adventitious buds are produced by isolated O. cinnamomea leaf primordia 2, 3 and 4; (2) the adaxial, adventitious buds are produced in addition to the leaf which bears them; (3) the adventitious buds are strictly foliar, not axial (cauline) in position or origin (Fig. 1); (4) expression of the adventitious, adaxial bud occurs only when the primordium is isolated from the rest of the shoot system (Fig. 2); (5) if one accepts the concept of adventitious budding in primordia 2, 3 and 4 in Osmunda, then it is equally valid to include primordia 5 through 9 as exhibiting the same response. From these investigations, it is apparent that determination of the primordium as a leaf primordium occurs

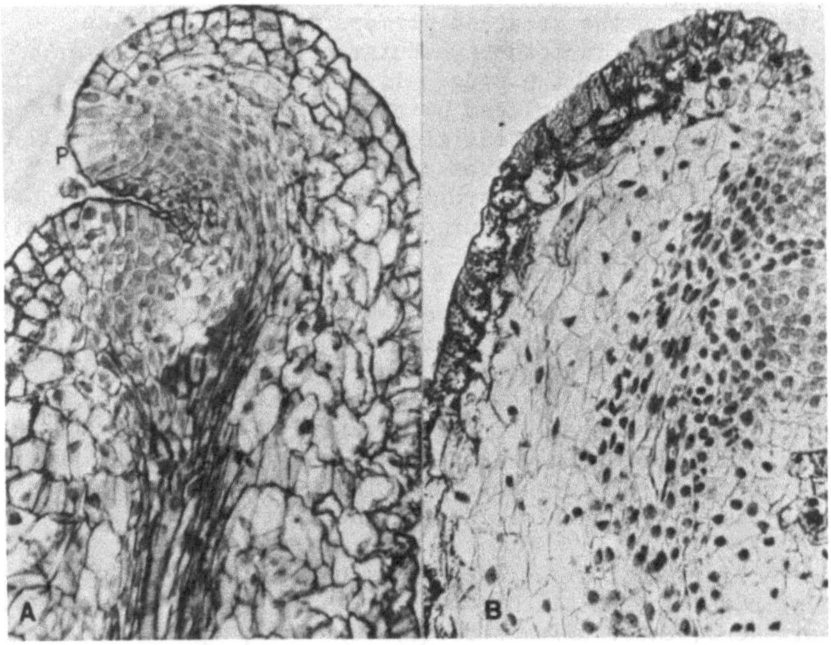

Fig. 1. Singly cultured leaf primordia of O. cinnamomea with SAM'. (A) The P_4 apex (P) is overtopping the SAM'; note the common vascular tissue. (B) P_2 primordial apex (P) with subjacent SAM' (S). Both primordia were cultured for a period of eight weeks. X 180. (Haight and Kuehnert[8]; reprinted with permission from the Canadian Journal of Botany.)

as early as the P_2 stage, and not as late as the P_{10} stage of development as earlier suggested (Steeves[20,21,22,23]; Kuehnert and Steeves[18]). From these investigations it was further demonstrated that the shoots developed by O. cinnamomea primordia 2 through 9 are not produced from the original leaf primordial apex as suggested by Steeves[22]. Rather, the observed "shoot" is produced basipetal to the primordial apex of the leaf (Fig. 1B) which bears the "shoot" (actually, adventitious bud)(Haight and Kuehnert[8]).

To emphasize the adventitious characteristic of the bud, histological studies reveal that in the region in which the adaxial, adventitious bud is expected to arise, relatively high levels of starch accumulate. As the bud emerges, the starch reserves decline. Starch is removed from all but fully differentiated cells of the leaf which supports the adventitious bud.

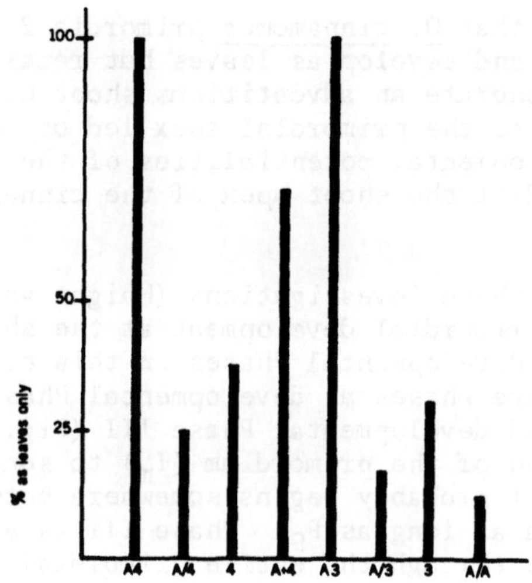

Fig. 2. The production of buds on the adaxial face of primordia cultured under various experimental conditions. Vertical axis: percentage of primordia expressed as leaves only; horizontal axis: experimental conditions. A4 = control plugs with SAM + P_4; A/4 = experimental plugs with P_4 isolated from the SAM, but both retained on the same plug; 4 = singly cultured P_4 ; A + 4 = SAM and P_4 cultured in physiological contact as a paired unit; A3, A/3, and 3 identical with A4, A/4, and 4 except that a P_3 is retained instead of a P_4. (Haight and Kuehnert[8]; reprinted with permission from the Canadian Journal of Botany.)

As in other systems (tobacco callus, Thorpe and Murashige[28]), it is significant that the main function of starch accumulation appears to be that it serves as a readily available source of energy for renewed meristematic activity within these systems.

The following points are also of significance. This phase of the study has demonstrated fully that there are two distinct developmental Phases in which primordia 2 through 9 will express. One of these developmental Phases takes the form of leaf + bud; the second takes the form of leaf only - the fully determined leaf primordium. Furthermore, the study strongly indicates that the Phase leaf + bud precedes the developmental Phase leaf only for younger primordia in the sequence express preferentially in the leaf + bud expression, while progressively older primordia express preferentially in the leaf only developmental Phase.

Emergence of a Three-phase Concept of Development

The discovery that O. cinnamomea primordia 2 through 9 are already determined and develop as leaves but retain the meristematic potential to generate an adventitious shoot bud adaxially and just basipetal to the primordial apex led us to a reinvestigation of the developmental potentialities of the youngest primordium produced at the shoot apex of the cinnamon fern, that is, P_1.

The data from these investigations (Haight and Kuehnert[9]) clearly show that primordial development at the shoot apex may be divided into three developmental Phases in this organism. We have designated these Phases as developmental Phase I, developmental Phase II, and developmental Phase III (Fig. 3). Phase I lasts from inception of the primordium (I_m) to some point in time during P_1. Phase II probably begins somewhere between I_m and I_1, and may be retained as long as P_9. Phase III is evident as early as P_1 and continues through the entire primordial sequence to include P_n. In nature, or under experimental conditions where physiological continuity between the primordium and the shoot apical meristem complex is maintained, O. cinnamomea primordial expression is Phase III expression (leaf only). However, if the primordia produced at the shoot apex are removed from certain external biological influences (specifically the shoot apical meristem and/or certain of the older primordia) terminal expression of the primordium may be either Phase I, Phase II, or Phase III depending upon the developmental Phase of the primordia at the time of their isolation. To elaborate: any primordium (P) or incipient primordium (I), under the normal influence of the shoot apical meristem complex, will express in developmental Phase III (leaf only). If the outside influence of the shoot apical meristem is removed during the time the primordium is passing through developmental Phase I, the primordium will express as shoot only. If the shoot apical meristem influence, and that of certain of the older primordia are removed while the primordium is passing through developmental Phase II, the primordium will express in the leaf + bud expression. However, if the primordium is not removed from the influence of the shoot apical meristem before it has passed on to developmental Phase III, it will express as a leaf only.

With regard to Phase I, when the influence is the shoot apical meristem, the probability that a Phase I primordium will develop as a leaf is 1.0. When the Phase I P_1 is removed from the external influence of the shoot apical meristem, the probability that the Phase I P_1 will develop instead as a shoot is also 1.0.

It is of further significance to recognize that only the youngest primordium of the cinnamon fern is able to express in all three developmental Phases. Though developmental Phase I is transient, it is a necessary Phase in the sequence which leads, ultimately, to the determination of the primordium as a leaf only. Furthermore, while the P_1 is in developmental Phase I it is quite misleading to refer to it as a "leaf" primordium. Rather, it should be regarded as a completely undetermined outgrowth of the shoot apex complex which has the capacity to develop either as a shoot or a leaf depending upon the external biological influences to which it is subjected.

Cell Cycle Characteristics of 21-day Old Third Emergent Primordia: Cell Subpopulations and Leaf Development

As stated above, it has been shown that the primordium initiated at the shoot apex of the cinnamon fern passes through

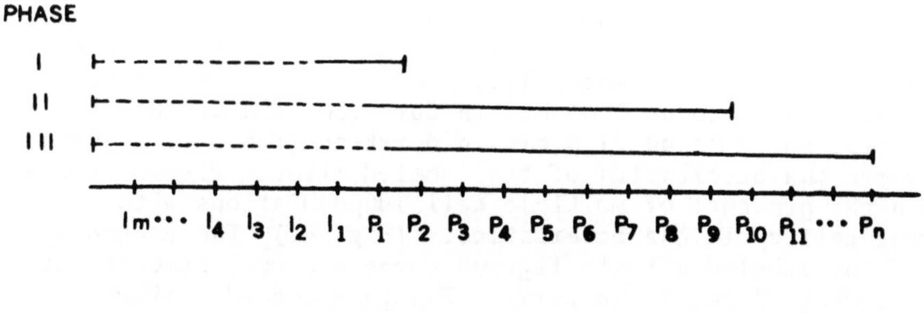

Fig. 3. Leaf determination in O. cinnamomea. Developmental Phases. (See text for further explanation.) (reprinted with permission from the Canadian Journal of Botany; Haight and Kuehnert[9].)

three developmental Phases, the final one of which is the adult, leaf only, Phase of expression. Of the three Phases, the intermediate Phase (Phase II), is the one which has received the bulk of our attention in the study of leaf development. As the intermediate Phase, it sits astride the "aging process" which proceeds from shoot only through to leaf only. Of the three Phases it is probably the most significant for it bears resemblance to the other two Phases flanking it in that it expresses as leaf + bud.

Third emergent primordia express preferentially in this intermediate Phase when excised and grown under the conditions of sterile nutrient culture; and whole body irradiation of cultured P_3's at a dose of 500 R x-rays will promote an increase in the expression of P_3's in this Phase (II - leaf + bud). In view of this, it appeared that P_3's offered a unique opportunity to study the peculiar characteristics of the leaf determination process through studying the unique cell cycle characteristics of the two distinctly different major cell populations of the third emergent primordium.

The labeled mitosis technique (Quastler and Sherman[19]) allows one to determine the duration of the cell cycle (CT) and its component parts: (G 1) the pre-synthetic interphase period, (S) the DNA synthesis period, (G 2) the post-synthetic interphase period, and (M) the period of mitosis. In plant systems, the labeled mitosis technique has been used extensively on root tip meristems (Wimber[34]). Determination of the cell cycle has been made both on whole root tips (Ames and Mitra[1]; Burholt and Van't Hof[3]; Van't Hof[29,30,31]; Van't Hof and Ying[33]; Wimber[35]) and on root sections (Clowes[4,5]; Thompson and Clowes[27]). Recently a report has been made of cell cycle characteristics in a shoot meristem (Jacqmard[10]).

Briefly stated, through the use of continuous and pulse labelling techniques our investigations of the cell cycle characteristics of excised, cultured 21-day old third emergent primordia (Phase II plus Phase III) show G 2 + 1/2M to be 6.75 hrs in duration, M to be 1.68 hrs in duration, minimum G 2 to be 5.91 hrs, and S to be 21.0 hrs in duration (Table I). Furthermore, curve characteristics of the labeled mitotic figures curve indicate the presence of multiple cell subpopulations with differing cell cycle characteristics. (Fig. 4.) The ascending slope of the labeled mitotic figures curve was much steeper than the descending slope of the curve. Though a second maximum emerged at 56 hours in the cycle time curve, the curve demonstrated sufficient irregularity in the vicinity of this second maximum to further suggest cell population heterogeneity. Finally, the curve characteristic did not allow determination of a G 1 interphase period, which further substantiates the probability of the existence of multiple cell subpopulations emerging within the system.

TABLE I, DURATION (HOURS) OF THE MITOTIC CYCLE AND ITS COMPONENT PHASES IN 21-DAY OLD P_3's OF O. CINNAMOMEA L.

G_2 + 1/2M	M	minimum G_2	S
6.75	1.68	5.91	21.0

Since the adaxial, adventitious bud emerges in P_3's cultured approximately 3 weeks (21 days), it is predicted that with the emergence of the bud, a change in the cell cycle characteristic will be expressed, and this will occur either as cell subpopulations with identical mitotic cycle durations, but differing period durations within the overall cycle, or cell subpopulations each with a different mitotic cycle time, and differing interphase period durations (Van't Hof and McMillan[32]).

Sources of an 'Active Principle'

From previous investigations (Kuehnert[13,14,15,16]) it has been demonstrated that certain of the older primordia (10's through 13's) are sites of production of a biologically active 'principle' which suppresses the leaf + bud expression forcing the developing primordium into the leaf only expression

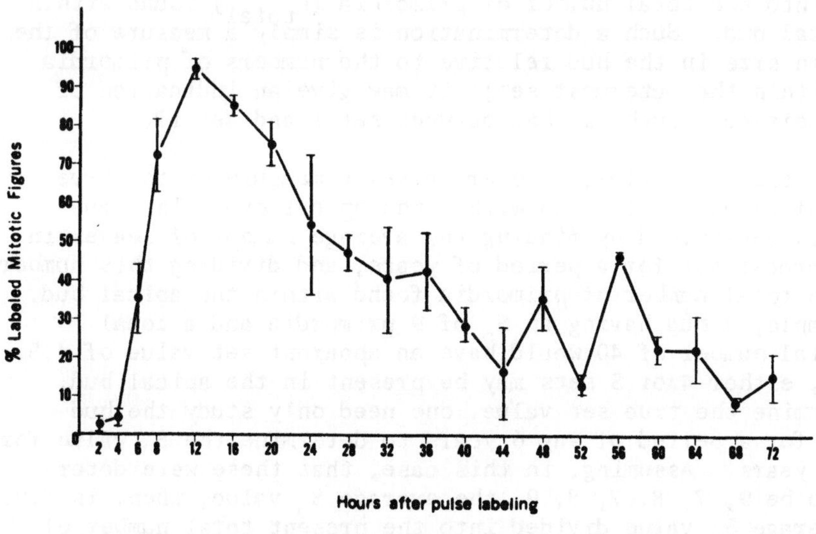

Fig. 4. Cell cycle characteristics of 21-day old P_3's of O. cinnamomea L. (See text for explanation.)

(Table II). From the investigations carried out by Haight and Kuehnert[8], it is suggested that the rhizome apical meristem (Fig. 2) is a second source of 'principle' which will suppress the leaf + bud Phase forcing the developing primordium to express, then, preferentially in the leaf only Phase of expression.

Other evidence comes from a series of experiments in which excised third emergent primordia of the cinnamon fern were grown, as pairs, in physiological contact with excised rhizome shoot apical meristem cones (Table II). Third emergent primordia expressed preferentially in the leaf only expression at a level equal to that characteristic of P_3's grown in physiological contact, as pairs, with certain of the older primordia (10's, 12's, 13's).

The Sets Characteristic

We have also determined a unique set value characteristic in our study of leaf development in this organism; and we recognize this characteristic to be intimately bound up with the developmental sequence for primordia initiated at the shoot apex, a sequence we have described in terms of Phases of development.

Two set values may be determined for O. cinnamomea. One of these is already in the literature (Steeves and Wetmore[26]; Steeves and Sussex[24]). We term this set value the apparent set value. The apparent set value is found by determining the total number of primordia in the outermost set (S_o), and dividing that number into the total number of primordia (P_{total}) found within the apical bud. Such a determination is simply a measure of the change in size in the bud relative to the numbers of primordia found within the outermost set; it may give an indication of sets interfaces, such as that between set I and set II.

The true set value, however, gives a measure of the true number of annual sets found within the apical bud. The true number is determined by finding the average number of leaves in the outermost set for a period of years, and dividing this number into the total number of primordia found within the apical bud. For example, a bud having an S_o of 9 primordia and a total primordial number of 40 would have an apparent set value of 4.5. However, either 4 or 5 sets may be present in the apical bud. To determine the true set value, one need only study the bud history for a period of say 6 years to determine the S_o value for those 6 years. Assuming, in this case, that these were determined to be 9, 7, 8, 7, 8, 9, the average S_o value, then, is 8.0. This average S_o value divided into the present total number of primordia (40) found in the apical bud will give the true number of annual sets (true set value) of 5.0.

TABLE II, FATES OF EXCISED P_3 PRIMORDIA CULTURED SINGLY, OR IN PHYSIOLOGICAL CONTACT AS PAIRS WITH OTHER P_3'S OR END OF SET I PRIMORDIA, OR EXCISED APICAL MERISTEM CONES.

Pairing conditions	Morphogenetic response		Leaves (%)
	shoots	leaves	
P_3's alone	7	3	30.00
P_3's with P_3's	26	16	38.00
P_3's with P_{12}'s	13	24	65.00
P_3's with P_{13}'s *	5	13	72.00 *
P_3's with P_{14}'s *	15	8	35.00 *
P_3's with apical meristem cones	4	16	80.00

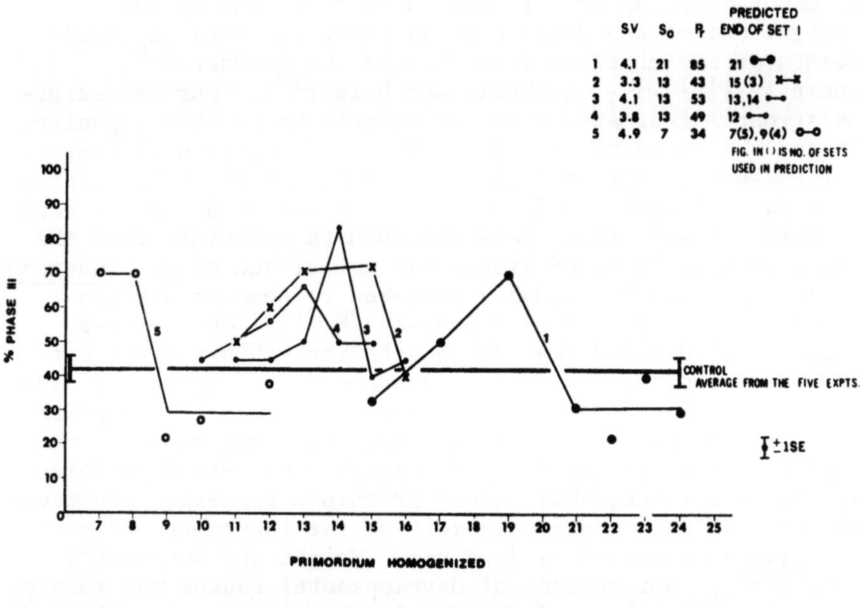

Fig. 5. O. cinnamomea P_3's assayed for preferential expression in developmental Phase III (leaf only). Control value obtained from P_3's grown on homogenate of P_3's. Experimental values obtained from P_3's grown on homogenate of predicted end of set I, beginning of set II primordia. Ratio of P_3:P_x maintained at 1:1. Standard error of the mean is indicated.

If one anticipated only 4.0 sets, or even 4.5 sets, based only upon information derived from apparent set value, and had not determined the true set value from bud history, the prediction of the sets interfaces could be in error by several plastochrons. This is of considerable significance when it is remembered that we recognize that certain, but not all, of the older primordia are intimately involved in the determination of the final expression of younger set I primordia (Kuehnert[13,14,15,16]). And our knowledge of this allows us to manipulate and study the details of leaf development. Determination of the true set value serves as a predictive tool, then, for determining sets interfaces, the value of which is shown by the fact that we now recognize that it is not a particular primordium, i.e., P_{13}, which is so influential in determining the final morphogenetic expression of younger primordia, for example P_3's, as might have been inferred (Kuehnert[15,16]). Rather we now realize that end of set primordia are responsible (Fig. 5).

DISCUSSION

The suggestion (Arber[2]) that the leaf exhibits an "urge to whole shoot-hood" appears to have been borne out by our investigations of the developmental potentialities of (leaf) primordia of the cinnamon fern (Haight and Kuehnert[8,9]; Kuehnert [13,14,15,16]; Kuehnert and Haight[17]). Our investigations strongly indicated that the primordium, in this organism, exhibited the capacity to express in two developmental Phases. One of these was expressed as leaf + bud, and the other expressed as leaf only (Haight and Kuehnert[8]). It was demonstrated that both Phases of expression were confined to primordia from the youngest set (set I) found within the apical bud of *O. cinnamomea*; and only primordia 2 through 9 appeared to demonstrate the observed plasticity (see also Steeves [20,21]). Our earlier studies also indicated that the plasticity exhibited was not random in character, but rather was part of a sequence of events, the leaf + bud characteristic preceding the leaf only expression (Haight and Kuehnert[8]). This was suggested from the fact that younger primordia of the group 2 through 9 expressed preferentially as leaf + bud, while older primordia expressed preferentially as leaf only; and primordia of the innermost set older than 9 always expressed as leaf only (Haight and Kuehnert[8]; Steeves[20,21]). Our concept of developmental Phases was reinforced from our investigations of the developmental potentialities of the youngest primordium produced at the shoot apex, P_1.

From our studies (Haight and Kuehnert[9]), we have learned that P_1 not only expresses in the two Phases already recognized, but it may also express as shoot only; and we term this Phase of expression developmental Phase I. This expression-Phase is observed to precede that of leaf + bud (developmental Phase II), and this in turn is observed to precede the leaf only expression (developmental Phase III) in this primordium. Thus, we see in this youngest primordium the embodiment of the so-called "urge to whole shoot-hood", for a primordium produced at the shoot apex, under the appropriate conditions, is observed to progress from shoot only through to leaf only potential; and equally important, determination of the primordium is an accomplished fact as early as the P_1 stage of development in O. cinnamomea. It is also significant that the sequence of observed developmental events is never expressed unless the primordium is isolated from the rest of the shoot system of which it is so integral a part. Of equal significance is the fact that we have demonstrated that the rest of the shoot system is intimately bound up with the control of morphogenetic events that occur in this region of great plasticity (in O. cinnamomea, the set I primordia). This is apparent from action of a well-documented 'active principle' (Haight and Kuehnert[8,9]; Kuehnert[13,14,15,16]) as well as other whole bud activities discussed earlier.

Not only have we been able to demonstrate this sequence of developmental events that progresses from totally uncommitted to partially committed to totally committed as leaf only development, but we also have been able to determine, reasonably accurately, the sources or sites of control. Repeatedly we have demonstrated that one of these sites is the shoot apical meristem, for isolated younger primordia grown in physiological contact with the shoot apical meristem (in pairs with isolated cones, isolated by microsurgical incisions, or not isolated) invariably express as leaf only primordia. Similarly also, we have demonstrated that primordia of the cinnamon fern which comprise the first set of (leaf) primordia are a second source of control, for, again, isolated younger primordia grown in physiological contact with other isolated older primordia express preferentially as leaf only primordia. Further, we have demonstrated that greatest control over the final morphogenetic expression of younger set I primordia is exerted by the oldest sibling primordia of these younger set I primordia; and preliminary data suggest that only oldest primordia of only set I are able to exert such strong influence over younger primordial development (Fig. 5).

Data accumulated to date suggest that in O. cinnamomea the plasticity of expression of a meristem-produced primordium is probably a set I phenomenon only. Further, the significant parts of the shoot system bound up in this control are also probably intimately associated with set I. Now that we have better defined the true developmental sequence of events in the (aging) determination of a primordium, we need to know, definitely, whether control of plasticity is only a set I phenomenon for this will better define the actual limits of the control sites and allow us to better characterize these in terms of 'active principle'. Further, we need to characterize 'active principle' both in size characteristics and in physical and chemical terms. Is 'active principle' no more than one of the known plant growth regulators? Or is 'active principle' a molecular complex, whose function is, principally, if not solely, to strip away the "youthful" characteristic from the developing primordium thus determining (aging) the primordium as (to) Phase III (leaf only) expression? If we can determine the outer limits of sites of production of 'active principle' we have better opportunity to determine whether 'active principle' of the two "boundary" interfaces (shoot apical meristem and oldest set I primordia) is similar or dissimilar from each of these sites.

It is fortuitous that whole body radiation techniques promote the partially committed expression (leaf + bud) in this system. It is this intermediate Phase of the observed developmental sequence that holds much promise for better understanding of the developmental sequence at those levels of organization where determination has true meaning, that is, at cell populational, cytological and biochemical levels. By using the labeled mitosis technique we will begin pursuit of the determinative phenomenon in those terms, approached through the cell cycle, with the view to determining those shifts at the cellular and subcellular level which convert non-leaf oriented cells to leaf-oriented cells at specific loci on the shoot apex. Thus we may have a fuller understanding of the phenomena associated with leaf determination and leaf development.

REFERENCES

1. Ames, I. H. and Mitra, J. (1966) The mitotic cycle time of Haplopappus gracilis root tip cells as measured with tritiated thymidine. Nucleus, 9:61.

2. Arber, A. (1950) The natural philosophy of plant form. Cambridge University Press.

3. Burholt, D. R. and Van't Hof, J. (1971) Quantitative thermal-induced changes in growth and cell population kinetics of Helianthus roots. Am. J. Bot., 58:386-393.

4. Clowes, F. A. L. (1961) Duration of the mitotic cycle in a meristem. J. Expt. Bot., 12:283-293.

5. Clowes, F. A. L. (1965) The duration of the G 1 phase of the mitotic cycle and its relation to radiosensitivity. New Phytol. 64:355-359.

6. Cutter, E. G. (1956) Experimental and analytical studies of pteridophytes. XXXIII. The experimental induction of buds from leaf primordia in *Dryopteris aristata* Druce. Ann. Botany (NS), 20:143-165.

7. Cutter, E. G. (1965) Recent experimental studies of the shoot apex and shoot morphogenesis. Botan. Rev., 31:7-113.

8. Haight, T. H. and Kuehnert, C. C. (1969) Developmental potentialities of leaf primordia of *Osmunda cinnamomea*. V. Toward greater understanding of the final morphogenetic expression of isolated set I cinnamon fern leaf primordia. Can. J. Botany, 47:481-488.

9. Haight, T. H. and Kuehnert, C. C. (1971) Developmental potentialities of leaf primordia of *Osmunda cinnamomea*. VI. The expression of P_1. Can. J. Botany, (in press).

10. Jacqmard, A. (1970) Duration of the mitotic cycle in the apical bud of *Rudbeckia bicolor*. New Phytol., 69: 269-271.

11. Jensen, W. A. (1962) Botanical Histochemistry. W. H. Freeman and Co., San Francisco.

12. Johansen, D. A. (1940) Plant Microtechnique. McGraw-Hill, New York.

13. Kuehnert, C. C. (1967) Developmental potentialities of leaf primordia of *Osmunda cinnamomea*. I. The influence of determined leaf primordia on undetermined leaf primordia. Can. J. Botany, 45:2109-2113.

14. Kuehnert, C. C. (1969a) Developmental potentialities of leaf primordia of *Osmunda cinnamomea*. II. Further studies of the influence of determined leaf primordia on undetermined leaf primordia. Can. J. Botany, 47:59-63.

15. Kuehnert, C. C. (1969b) Developmental potentialities of leaf primordia of *Osmunda cinnamomea*. III. Studies of the effects of homogenized, determined leaf primordia on expression-potential of undetermined leaf primordia. Can. J. Botany, 47:65-68.

16. Kuehnert, C. C. (1969c) Developmental potentialities of leaf primordia of Osmunda cinnamomea. IV. Expression potential of undetermined primordia separated by a barrier membrane from undetermined or determined primordia. Can. J. Botany, 47:69-72.

17. Kuehnert, C. C. and Haight, T. H. (1969) The shoot forming capacity of cinnamon fern leaf primordia: A reevaluation. (XI International Botanical Congress; Abstracts, p. 117).

18. Kuehnert, C. C. and Steeves, T. A. (1962) Capacity of fragments of leaf primordia to produce whole leaves. Nature, 196:187-189.

19. Quastler, H. and Sherman, F. G. (1959) Cell population kinetics in the intestinal epithelium of the mouse. Exptl. Cell Res., 17:420-438.

20. Steeves, T. A. (1961a) The development of leaves in sterile nutrient culture. In: Recent Advances in Botany, 1:823-827. Univ. of Toronto Press.

21. Steeves, T. A. (1961b) A study of the developmental potentialities of excised leaf primordia in sterile culture. Phytomorphology, 11:346-359.

22. Steeves, T. A. (1962) Morphogenesis in isolated fern leaves. In: Regeneration, (ed.) D. Rudnick (20th Growth Symp.), 117-151, Ronald Press, New York.

23. Steeves, T. A. (1966) On the determination of leaf primordia in ferns. In: Trends in Plant Morphogenesis. Edited by: E. G. Cutter. Longmans, Green and Co. Ltd., London. pp. 200-219.

24. Steeves, T. A. and Sussex, I. M. (1957) Studies on the development of excised leaves in sterile culture. Am. J. Bot., 44:665-673.

25. Steeves, T. A., Sussex, I. M. and Partanen, C. R. (1955) In vitro studies on abnormal growth of prothalli of the bracken fern. Am. J. Bot., 42:232-245.

26. Steeves, T. A. and Wetmore, R. H. (1953) Morphogenetic studies on Osmunda cinnamomea L. Some aspects of the general morphology. Phytomorphology, 3:339-354.

27. Thompson, J. and Clowes, F. A. L. (1968) The quiescent centre and rates of mitosis in the root meristem of Allium sativum. Ann. Bot. (NS), 32:1-13.

28. Thorpe, T. A. and Murashige, T. (1970) Some histochemical changes underlying shoot initiation in tobacco callus cultures. Can. J. Botany, 48:277-285.

29. Van't Hof, J. (1965) Relationships between mitotic cycle duration, S period duration and the average rate of DNA synthesis in the root meristem cells of several plants. Exptl. Cell. Res., 39:48.

30. Van't Hof, J. (1967) Studies on the relationships between cell population and growth kinetics of root meristems. Exptl. Cell Res., 46:335-347.

31. Van't Hof, J. (1968) Experimental procedures for measuring cell population kinetic parameters in plant root meristems. In: Methods in Cell Physiology, Vol. III, pp. 95-117. D. M. Prescott (ed.), Academic Press, New York.

32. Van't Hof, J. and McMillan, B. (1969) Cell population kinetics in callus tissues of cultured pea root segments. Am. J. Bot., 56:42-51.

33. Van't Hof, J. and Ying, H. K. (1964) Relationship between the duration of the mitotic cycle, the rate of cell production and the rate of growth of Pisum roots at different temperatures. Cytologia, 29:399-406.

34. Wimber, D. E. (1960) Duration of the mitotic cycle in Tradescantia paludosa root tips as measured with H^3-thymidine. Am. J. Bot., 47:828.

35. Wimber, D. E. (1966) Prolongation of the cell cycle in Tradescantia root tips by continuous gamma irradiation. Exptl. Cell Res., 42:296-301.

*It is a pleasure to acknowledge the collaboration of Dr. T. H. Haight, Mr. E. H. Jordan, and Dr. Jack Van't Hof in some of the experimental work published here for the first time. The assistance of Mrs. Bertha M. Kuehnert in preparing some of the text figures is gratefully acknowledged. Financial support was provided by the United States Atomic Energy Commission, Division of Biology and Medicine, Biology Branch.

-DISCUSSION-

DAVIDSON - Do you think this switch in cycle time is gradual or sudden?

KUEHNERT - Believing leaf and bud to have their own characteristic cycle times I will have to work through the developmental sequence to determine this.

SMITH - I was wondering if length of culture of these buds influences their ultimate expression of differentiation?

KUEHNERT - No, there is only a limited period of culture before they express themselves which would not be long enough to alter their characteristic commitment.

CUTTER - In *Dryopteris* (by microsurgical isolation) you can alter the expression of primordia by inducing temporary regression, and then allowing them to redevelop at a later time.

TORREY - If you are interested in isolating the two cell types you have here, had you thought about using the technique of protoplasts? Having gotten these out, they might form new walls, enabling you to start off with single cells.

KUEHNERT - I'm not sure that the differences in the cells from the two populations are such that they would persist in culture.

BALL - You imply that a certain number of leaves show irreversible differentiation from a very early stage; and it seems to me a very strong assertion for plants.

MECHANISMS OF PERCEPTION AND PATTERNS OF ORGANISATION IN ROOT CAPS

BARRIE JUNIPER

School of Botany, The University of Oxford

South Parks Road, Oxford OX1 3RA, England

INTRODUCTION

In Guttenberg's review[5] of root caps, their functions are listed as protecting the meristem from damage and acting as a drilling tissue through the earth. Both of these functions are served by the outer two or three layers of cells forming the cap periphery; a third function, the perception of gravity (Juniper and French[12]) is provided by the cells of the cap columella or core cells (Fig. 1)

The populations of cells that serve as a protective layer and serve as a gravity perceiving system are, as we shall see, quite distinct in their fine structure although derived from a common meristem. The change from core cell type (C.C. in Fig. 1) to peripheral cell type (P.C.) takes place abruptly and unambiguously in both space and time.

MATERIALS and METHODS

Seeds of Zea mays, varieties Kelvedon 33 and Golden Bantam, Pisum sativum varieties Alaska and Weitor, and Vicia faba were washed overnight in running water, germinates on filter paper and, when the primary root was about 2-3 cm long, were fixed and stained in the following ways. a) In 2% glutaraldehyde in cacodylate buffer at room temperature for two hours, followed by two hours in 2% OsO_4 also in cacodylate buffer either in crushed ice or at room temperature b) in 2% glutaraldehyde for 2 hours in 2% $KMnO_4$, all at room temperature c) Directly into 2% $KMnO_4$ for 2 hours at room temperature where a rapid fixation was wanted.

Some roots were taken, while still growing and clipped unbroken into a perspex (lucite) frame or grown on in 2 m.m. glass tubes to give them a specific orientation for a given time before fixation.

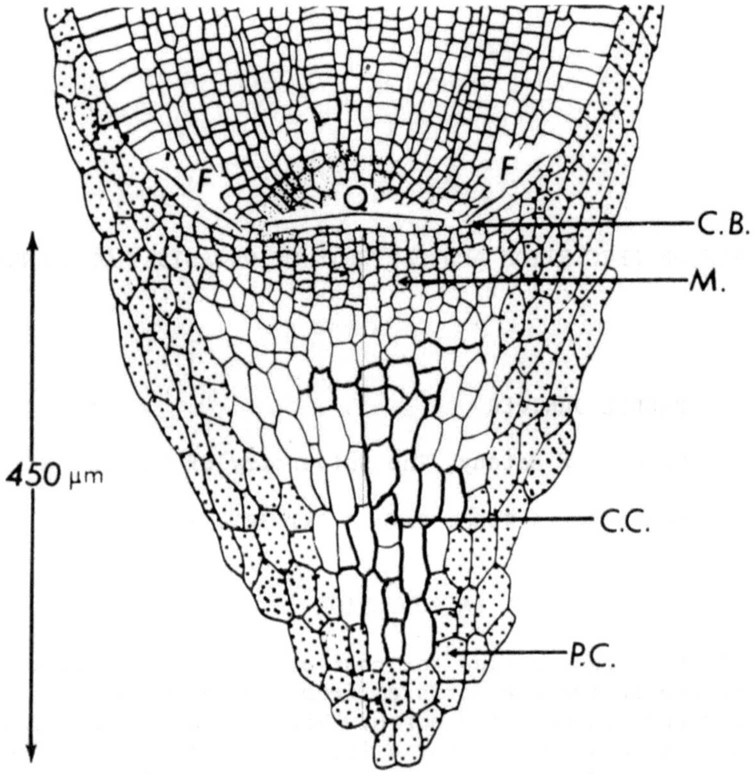

Fig 1.

Fig. 1. Outline drawing of the cells of the tip of the primary root of maize (Zea). Reproduced by kind permission of Springer Verlag from Planta (Berl.) 95, 315, 1970

The level of the fixative was raised around the still-growing roots clamped in their determined orientation; the roots were not moved in the fixative.

After fixation the roots were embedded in Epikote 812 or Coulter Epon* resin, sectioned with glass or diamond knives on a Cambridge Huxley or Reicherts OMU2 microtome and examined in an AEI EM6, Philips 200 or Zeiss EM9 electron microscope. For some of the measurements large numbers of individual micrographs were taken at x 1500 of 2-3 adjacent serial sections and built up into complete mosaics of the root cap or tip, to give final magnifications of x 3,800.

* Coulter, H. D. (1967) J. Ultrastruct. Res. 20: 346.

RESULTS

General Observations

If we summarize briefly the fine structural characteristics of the core cells of grasses, they have fat, sedimenting amyloplasts, peripherally distributed, rough-surfaced endoplasmic reticulum (ER) (see Fig. 3 (1)), no vacuoles, relatively thin permanganophilic walls and relatively inactive Golgi bodies. Both the amyloplasts and the ER, at least in grass root caps, will respond to a gravitational stimulus in predictable ways. The amyloplasts of peripheral cells are shrunken, may finally lose all their starch and neither they nor the ER respond to a gravitational stimulus. The ER is randomly distributed within the cell, often in the form of clumps or whorls, and scattered here and there are small, irregularly-shaped vacuoles (Fig. 7 in Juniper and French [12]) The ER is not disturbed by nor recovers from a gravitational disturbance. The walls are swollen, do not respond to permanganate stains and the plasmalemma lining the walls of these cells is wrinkled and curved suggesting vigorous reverse pinocytotic activity. The Golgi bodies do not increase in number (Juniper and Clowes [9]), but increase in size and produce large numbers of swollen vesicles. These hypertrophied vesicles migrate to the plasmalemma and their contents finally pass through the plasmalemma, through the wall and then join the mucus coating the outside of the root cap. This whole process of vigorous reverse pinocytosis has been well documented by Mollenhauer <u>et al</u>, Northcote and Pickett-Heaps [16], Juniper and Roberts [10].

All the changes listed above take place more or less simultaneously. There is, however, a continuing loss of starch from the amyloplasts, a progressive weakening of the wall structure and growth of vacuoles. Finally these cells spontaneously detach from the cap margin and drift away into the mucus, semi-vacuolate, virtually protoplasts, but still apparently alive. The abrasion of soil particles is not necessary for the removal of peripheral cap cells, their loss is organised and co-ordinated and will take place even when the roots are grown on a soft agar surface.

The Mechanism of Perception in Peripheral Cap Cells.

We have recently observed an additional feature of these peripheral cells, which both reinforces the idea that they are not just a suicide squad, but a highly organised tissue, and may have application to other tissue development.

As the discharge of vesicles from the Golgi gets under way, Northcote and Pickett-Heaps[16] have shown that the Golgi-derived material accumulates briefly between the plasmalemma and the wall. We have been able to show, by building complete mosaics of root caps, that this accumulation almost always takes place between the plasmalemma and what is, or what will be, the outer tangential wall (Fig. 2). Fig. 2 is an outline drawing of an electron micrographic mosaic of a root cap in which the accumulated Golgi

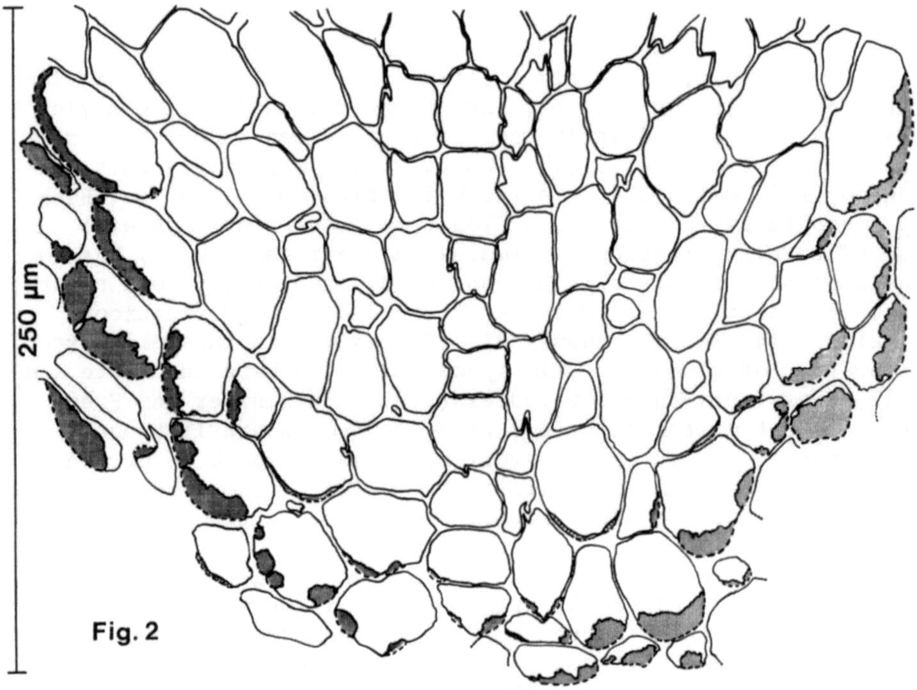

Fig. 2 Outline drawing of plasmamembranes from an electron-micrograph mosaic of the cells of the distal tip of a primary root of maize. The dotted areas represent Golgi-secreted material accumulated between the plasmamembrane and the wall.

material is distinguished by dotting. The solid line represents the plasmalemma and the dotted line the boundary between the accumulated Golgi material and the wall proper. Some parts of extreme peripheral cells were missing from this mosaic and their margins are shown incomplete.

Thus peripheral cells, at least three cells in from the extreme margin, possess a perception mechanism able to predict which wall will finally constitute the outer tangential wall. This perception and selective migration system works too, at least in maize, regardless of the orientation of the root, i.e. roots grown free from pressure in the normal, vertical position or pointing upwards or roots grown resting on moist filter show the same phenomenon.

What is this mechanism? A symplastic mechanism was sought for. The distribution of the Golgi bodies themselves is, from a study of our micrographs, random within each cell. The path of the Golgi vesicles to and the point of their fusion with the plasmalemma is not easy to investigate on static electron micrographs. However in so far as this can be ascertained from our micrographs and from an examination of the path of radioactive-labelled material in the work by Northcote and Pickett-Heaps[17] it appears also to be random. Microtubules are rare in cap cells, but where they are present they appear to have no relationship to the movement of Golgi vesicles to the wall. No obvious polarity of any organelle or its derivative within these cells exists.

The simplest explanation now appears to be that Golgi-derived vesicles migrate randomly to, and fuse with the plasmalemma, all over its surface. They then pass their contents through the plasmalemma. The soft, plastic pectin-rich material (Northcote and Rickett-Heaps[17])now lies between the plasmalemma and the wall. It begins to migrate, still in the region between the plasmalemma and the wall, until it reaches and accumulates in the area of the wall which will subsequently become the outer tangential surface. The force behind such a migration may be the unequal pressures developed, as a result of the architecture of the root, on the different wall surfaces. As each individual cell moves to the margin of the cap, the new outer tangential wall will become exposed on the surface and the Golgi-derived material now migrates through the softened wall to join the mucus on the outside.

Such a selective migration of pectin-rich material might, for example, be able to explain the accumulation and augmentation of the pectin layer which is commonly found between the cuticle and epidermal cell wall of the aerial parts of the plants.

The Mechanism of Perception in the Core Cells

The outer three layers of cells have one type of perception mechanism; the core cells have another. In fact, as Schachar[18] has shown, when caps are removed from roots (decapping), that portion of the root cap which needs to be regenerated for the root to be able to perceive gravity approximately corresponds to the core zone. Thus the peripheral cap cells are not involved in gravi-perception, but are involved in a perception system of their own.

Although the normal gravity perception system of the caps of grasses requires almost, but not quite, all the longitudinal development of the cap, as Gibbons, and Wilkins[3] have elegantly shown, a 'message' may be fed back to the elongating zone of the root by only half a root cap provided that it is cut in the longitudinal plane. Gibbons andWilkins roots, regardless of their orientation, always curved in the direction of the half cap. Thus only half of a core cell's population is necessary,but its longitudinal development must be complete. These observations suggest that little if any transverse migration within the cap takes place.

What features of core cells may be significant in determining the direction of a gravitational stimulus? It may be relevant,in view of Gibbons and Wilkins observations above, that the plasmodesmata (Table I) are most abundant on the transverse walls.

TABLE I
Average number of plasmodesmate per 100 μm^2

Regions +	Transverse walls	Longitudinal walls	Nos. of roots
P.C.	513 ± 123*	45 ± 6	3
C.C.	772 ± 288	75 ± 14	4
M.	1487 ± 208	530 ± 108	7
C.B. region Q	576 ± 137	-	7
C.B. regions F	-	152	1
Quiescent centre	950 ± 155	990 ± 155	6

* standard deviation: (data adapted from Juniper & Barlow[11])
\+ see Fig. 1 for explanation of abbreviations.

Thus if plasmodesmata are involved in the transmission of a stimulus an excellent longitudinal system exists, but transverse movement, at least within the cap, is negligible. Such a pattern may explain the results of Schachar, and Gibbons and Wilkins. Only half, or possibly less than half a core cell population need be present, but this half population must comprise continuous and complete files of core cells. Moreover, since Schachar's experiments on decapping and replacing maize root caps failed to restore the gravi-perceptive system it looks as though, at least within the cap, plasmodesmatal continuity is necessary for the transmission of the stimulus.

Within the core cells, as we have seen, the amyloplasts sediment rapidly on the receipt of a gravitational stimulus. This sedimentation pattern can accurately be correlated with the receipt of a gravitational stimulus and these amyloplasts are widely believed to act as statoliths (Audus[2]; Iversen[7] and Iversen et al[8]). However others doubt the connection (Anker[1]). The movement of the amyloplasts is facilitated both by their large size (up to 5 µm across their longest axis), their small numbers (there are about 35 in each of the core cells of *Zea*), the relative absence of vacuoles in the core cells, the apparent low density of the core cell cytoplasm and the peripheral nature of the ER (Juniper and French[12]).

In *Zea* and *Vicia* core cells only one other cell component, the ER, has been observed to move consistently when the cell perceives a gravitational stimulus. The pattern of the ER in undisturbed core cells is predictable.

As Fig. 3 (1) shows the ER normally lies parallel to the walls of the cell and, to a lesser extent, parallel to the nuclear membrane. The ratios given were measured directly from standard

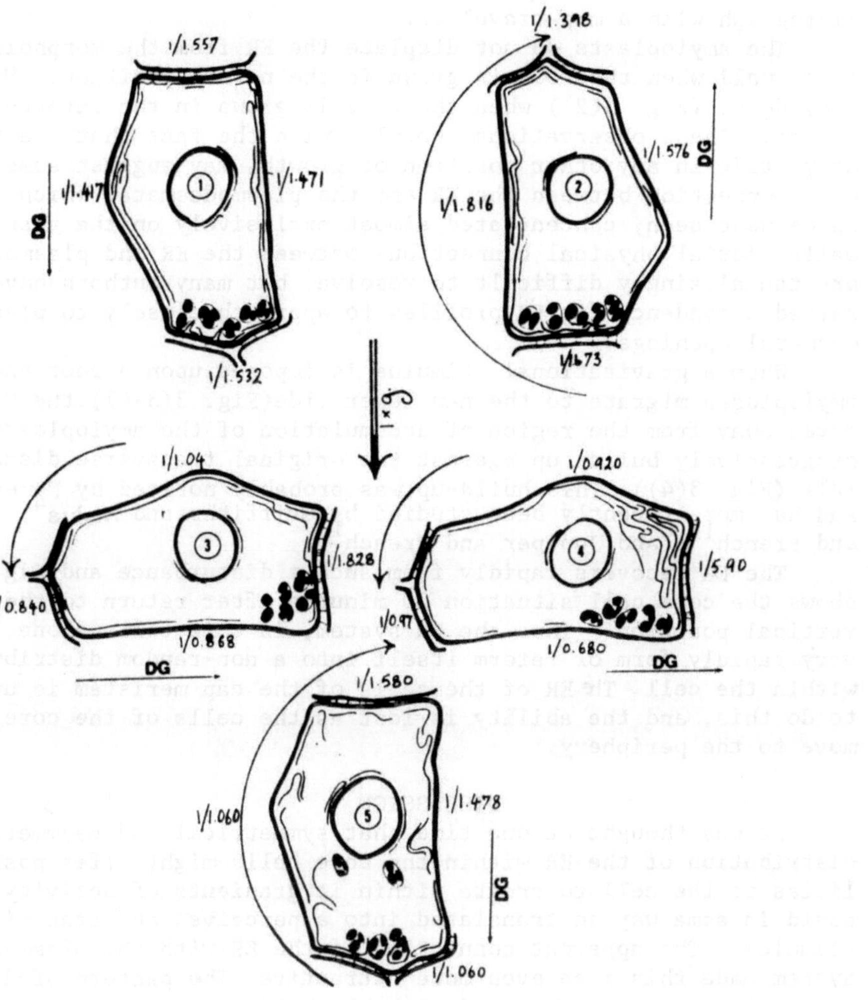

Fig. 3 The effect of different gravitational disturbances on the amyloplasts and ER of the core cells of Zea mays. The figures against each wall represent ratios of the length of the plasmalemma to the length of the ER parallel to that region of the plasmalemma. DG = direction of growth: 1 x g = direction of gravity.

Fig. 3(1) The normal appearance of a core cell after 2 hours in the vertical position.

Fig. 3(2) The appearance of a similar core cell 2 hours after 180° inversion.

Fig. 3(3) After a 10 minute horizontal stimulus.

Fig. 3(4) After a 20 minute horizontal stimulus.

Fig. 3(5) After 2 hours irregular disturbance and 20 minutes in the normal vertical position.

micrograph with a map-traveller.

The amyloplasts do not displace the ER from the morphologically lower wall when the root is grown in the normal position. Nor do they do so (Fig. 3(2)) when the root is grown in the inverted position. These observations, coupled with the fact that the ER is assymetric in any other position of growth, may suggest some form of a connection between the ER and the plasmodesmata, which are, as we have seen, concentrated almost exclusively on the transverse walls. Actual physical connections between the ER and plasmodesmata are tantalizingly difficult to resolve, but many authors have observed a tendency for ER profiles to approach closely to plasmodesmatal openings.

When a gravitational stimulus is imposed upon a root cap, the amyloplasts migrate to the new lower side(Fig. 3(3-4)),the ER moves away from the region of accumulation of the amyloplasts and progressively builds up against the original transverse distal wall (Fig. 3(4)). This build-up was probably noticed by Nemec[15 & 16] and has more recently been studied by Griffiths and Audus[4], Juniper and French[12], and Juniper and French[13].

The ER recovers rapidly from such a disturbance and Fig. 3(5) shows the core cell situation 20 minutes after return to the normal vertical position. Thus the ER system, in core cells alone, can very rapidly form or reform itself into a non-random distribution within the cell. The ER of the cells of the cap meristem is unable to do this, and the ability is lost as the cells of the core region move to the periphery.

DISCUSSION

It was thought at one time that symmetrical and asymmetrical distribution of the ER within the core cells might offer possibilities to the cell to create within it gradients of activity, which could in some way be translated into a perceived and transmitted stimulus. The apparent connection of the ER with the plasmodesmatal system made this idea even more attractive. The pattern of distribution that we have observed probably holds for many grasses and some other plants e.g. *Vicia faba*. However, as Grundwag and Juniper have discovered, some other plants, e.g. *Pisum sativum* vars.Weitor and Alaska have no ER system comparable to that observed in *Zea* and *Vicia*. The ER lies in small clumps or whorls at the distal end of the core cells and is not redistributed by a gravitational disturbance.

These observations do not exclude the ER entirely from some role in the gravi-perceptive mechanism, but its role is no longer as obvious as was first thought.

If not a gravi-perceptive mechanism what role does the ER system play? Although not yet confirmed it is possible that these membranes may be responsible for forming one of the transient wall components of the core cells of certain species.

The ER system cannot universally be implicated as an intermediate in a perceptive system. Nor can we detect any other change in the pattern of distribution of cell components apart from the amyloplasts, which might act as an intermediate. We are, therefore, reduced to assuming that, if the amyloplasts are statoliths, and the evidence is by now almost overwhelming, they must act in some direct way, either perhaps by interfering selectively with the flow of materials through the plasmodesmata or perhaps by altering the permeability of plasmalemma upon which they may come to lie.

CONCLUSIONS

We have therefore demonstrated that the root cap, a very small part of the whole root tip, possesses two integrated perception systems, separate, but with a common origin. The outer system, the controlled discharge of mucus, is dependent upon a co-ordinated loss of starch from the amyloplasts, the stimulation of the Golgi bodies, the disorganisation of the ER so as to provide no impedance to the movement of vesicles to the plasmalemma and a general weakening and increase of permeability of the cell wall. Given the capacity of pectin-rich polysaccharides to flow under pressure it is not difficult to envisage a mechanism whereby the directed accumulation and discharge is achieved.

At the centre lie the core cells, almost certainly the site of perception of gravity in grass roots and probably, although this is more difficult to prove, in other plants as well.

These cells, as we have shown, are exceptional in being almost non-vacuolate, have a co-ordinated organelle population with large numbers of small mitochondria and Golgi, and small numbers of large plastids. They accumulate within their plastids, as they move out of the meristem, large amounts of starch, retain it tenaciously (Audus[2]) and only discharge it as they change their roles in the peripheral cells. They are linked, particularly through their transverse walls, by large numbers of plasmodesmata, none of which are occluded during the life of the cell, a regular feature of other cell development, but persist until they are lost at the edge of the root.

REFERENCES

1. Anker, L. (1968) On gravi-sensitivity in plants. Acta bot. néerl. 17: 385-389.

2. Audus, L. J. (1962) The mechanism of the perception of gravity by plants. in 'Biological Receptor Mechanisms'. Symp. Soc. exp. Biol. 16: 197-226.

3. Gibbons, G.S.B. and Wilkins, M.B. (1970) Growth inhibitor production by root caps in relation to a geotropic response. Nature (Lond.) 226: 558-559.

4. Griffiths, H. J. and Andus, L. J. (1964) Organelle distribution in the statocyte cells of the root tip of *Vicia faba* in relation to a geotropic stimulation. New Phytol. 63: 319-333.

5. Guttenberg, H. (1968) Der primäre Bau der Angiospermenwurzel. Handbuch der Pflanzenanatomie, Bd. 8, Teil 5, S. 138-141. Berlin: Gebrüder Borntraeger.

6. Grundwag, M. and Juniper B. E. (1971) unpublished observations on the fine structure of ageotropic mutants of pea (*Pisum sativum*).

7. Iversen, T. H. (1969) Elimination of geotropic responsiveness in roots of cress (*Lepidium sativum*) by removal of statolith starch. Physiol. Plantarum (Cph.) 22: 1251-1262.

8. Iversen, T. H., Pedersen, K., and Larsen, P. (1968) Movement of amyloplasts in the root cap cells of geotropically sensitive roots. Physiol. Plantarum (Cph.) 22: 1251-1262.

9. Juniper, B. E. and Clowes, F. A. L. (1965) Cytoplasmic organelles and cell growth. Nature (Lond.) 208: 864-865.

10. Juniper, B. E. and Roberts, R. M. (1966) Polysaccharide synthesis and the fine structure of root cap cells. J. roy. micr. Soc. 85: 63-72.

11. Juniper B. E. and Barlow, P. W. (1969). The distribution of plasmodesmata in the root tip of maize. Planta (Berl.) 89: 352-360.

12. Juniper, B. E. and French, A. (1970) The fine structure of the cells that perceive gravity in the root tip of maize. Planta (Berl.) 95: 314-329. Berlin-Heidelberg-New York: Springer.

13. Juniper, B. E. and French, A. (1972) The distribution and redistribution of ER in root tips. (in preparation)

14. Mollenhauer, H. H., Whaley, W. G., Leech, J. H. (1961) A function of the Golgi apparatus in outer root cap cells. J. Ultrastruct. Res. 5: 193-200.

15. Němec, B. (1900) Über die Art der Wahrnehmung des Schwerkraftreizes bei den Pflanzen. Ber. dtsch. bot. Ges. 18: 241-245.

16. Němec, B. (1902) Die Perception des Schwerkraftreizes bei den Pflanzen. Ber. dtsch. bot. Ges. 20: 339-354.

17. Northcote, D. H. and Pickett-Heaps, J. D. (1966) A function of the Golgi apparatus in polysaccharide synthesis and transport in the root-cap cells of wheat. Biochem. J. 98: 159-167.

18. Schachar, B. Landau (1967) The root cap and its significance in graviperception. Ph.D. thesis of the University of London, Bedford College.

ACKNOWLEDGEMENTS

I would like to thank Mrs. Anna French, my research assistant, for her help in the accumulation of much of these data, Dr. W. M. Laetsch of the Department of Botany, Berkeley, California, for his hospitality and for the provision of E.M. facilities in the later phases of this work, and Mrs. Marianne Spitzer for her help in the preparation of the manuscript.

-DISCUSSION-

FOSKET - Is *Allium* still considered to be a problem for the statolith theory?

JUNIPER - No, the so-called exceptions are not really exceptions if one looks at other possible sedimenting bodies.

BALL - Iverson, *et al* have shown that while statoliths occur, very few move in response to gravity (Phys. Pl., 21:811, 1968). They cast doubt upon the efficacy of such a small number and suggested an alternative mechanism.

JUNIPER - Yes, but in root tip cells which I have examined only those cells are responsible for gravity perception which have such a system.

CONGER, A. - Is there elongation in the root cap?

JUNIPER - That's correct.

FOSKET - Can a core cell become a peripheral cell?

JUNIPER - As far as I have been able to determine most core cells do in fact become peripheral cells.

D'AMATO - Miss S. Avanzi has shown with root caps that the peripheral cells incorporate large amounts of tritiated amino acids (unpublished data). Do you feel this is related to polysaccharide synthesis?

JUNIPER - Core cells are of course very active metabolically, but the peripheral cells exist principally upon their reserves; they seem to use up their starch reserve from their own amyloplasts. Northcote and Pickett-Heaps have demonstrated that they take up H^3 glucose and galactose (Biochem. J., 98:159-167, 1966). Could the apparent amino acid uptake by the peripheral cells be an expression of a diffusion process?

D'AMATO - I'm not certain.

SAVAGE - Does the mucilaginous material from the peripheral cells swell on those sides of the cell in contact with water?

JUNIPER - For all practical purposes, all cells are bathed in water, but the outer cell walls also are breaking down spontaneously. This probably aids the Golgi in secreting the mucilaginous material.

BALL - There appears to be a feedback system between the root cap and the apex in that when you remove the root cap the remaining portion of the apical meristem is strongly modified: it goes through a regenerative phenomenon which upon completion of a new root cap meristematic system shuts off.

JUNIPER - Yes. It would appear that the quiescent center depends upon the presence at least of a root cap meristem.

VAN'T HOF - Is the "quiescence" of the quiescent center brought about by internal hypoxic conditions?

JUNIPER - No.

TORREY - The metabolic activity of root segments has been shown to be quite high.

CONGER, A. - Ebert and Howard showed in roots a pumping mechanism for oxygen which provided oxygen in amounts far greater than would be expected by simple diffusion (Les Peroxydes Organiques in Radiobiologie, Pergamon Press, London. 1958).

DAVIDSON - Alma Howard has shown with oxygen[15] that it was present at the apex of the root within seconds of its application at the plant top. I think it is safe to assume that the quiescent center is not hypoxic (Les Peroxydes Organiques en Radiobiologie, Pergamon Press, London. 1958).

TORREY - In *Convovulus* there are intercellular spaces which reach the apical meristem.

JUNIPER - On a simple diffusion basis it would be extremely difficult to explain why a cell of the quiescent center is hypoxic and the root cap meristem cell (which is extremely active mitotically) is aerobic.

BALL - Is the material at the root cap cell wall interfaces pectin in nature? If so, this might allow them to slide over one another.

JUNIPER - Yes, it is pectic in nature; however, I have no evidence that the cells slip by one another. Slippage would require a reforming of plasmodesmata. The growth appears to be a simple progression of cells from the central zone to the periphery. A plot of the number of plasmodesmata against the approximate length of elongation fits a normal distribution pattern. However, a sliding mechanism would require a different pattern and hence a different distribution. This we don't see.

DAVIDSON - The cells of the root cap undergo an elongation *and* in the absence of the root cap the cells in the region of elongation of the root continue to elongate. This suggests two controlling mechanisms -- one in the root cap and one in the root.

CUTTER - Gibbons and Wilkins have, however, shown that when half of the root cap is removed root elongation is inhibited on the side with the root cap remaining (Nature, 226:558-559, 1970).

JUNIPER - Gibbons and Wilkins have also shown that when a longitudinal half of the root cap is removed the root always elongates toward the remaining half regardless of geotropic position (Nature, 226:558-559, 1970). This is difficult to reconcile with Schachar's observations where the rate of elongation remains the same in the absence of a root cap (Ph.D. Thesis, University of London, 1967).

HABER - Dr. Davidson's comment about controlling mechanisms is an intriguing one. Perhaps removal of the root cap results in two phenomena: (a) a removal of an inhibition and (b) an inciting of some injury. If these two phenomena balance each other, a steady rate of elongation would be the net result.

THE CONTROL OF CELL PROLIFERATION WITHIN ROOT MERISTEMS

F.A.L. CLOWES

Botany School, Oxford University

INTRODUCTION

It is useful for some purposes to consider roots as having a subdistal meristem extruding cells towards the base to lengthen the root and towards the tip to form the root cap. No meristem, however, is as simple as this. A complex spatial distribution of rates of cell growth and cell division is necessary because the cells are generated in two opposing directions and are smallest near the boundary between the cap and the rest of the root. In the meristem cells are arranged differently in different species and so the pattern of rates of cell growth and division differs from species to species. A morphologist can in fact predict what the relative rates of growth and mitosis in different regions of the meristem will be by analysing the arrangement of cells. In the roots of Zea, for example, it was possible to predict from the cell pattern that there is a quiescent centre to the meristem where rates of cell growth and division are negligible compared with those of the surrounding cells[4]. This prediction has been confirmed by labelling dividing cells and by direct measurements of rates of mitosis. In other species with different arrangements of cells there is also a quiescent centre to the meristem. Its shape may also be predicted from the cell pattern, but, in some species, more extensive knowledge of how meristems operate is needed before its existence can be deduced with certainty. The "open" type of meristem such as Vicia or Pisum possess has a quiescent centre which is shaped like a disc or a concavo-convex lens. Its distal surface may, in some species, become active for brief periods in giving rise to new sets of cap initials during certain stages of development. In spite of this its mean rate of mitosis is still about one tenth of that of the surrounding cells and the distal margin of the quiescent centre has at most times a visibly distinct type of cell wall separating it from the cap initials[10].

The cells of the quiescent centre are not inherently quiescent although they have many characters different from their neighbours. They have the lowest rate of DNA synthesis, the lowest average DNA content, the lowest RNA per cell, the smallest nucleoli with the least RNA, the lowest rates of RNA and protein synthesis, the fewest ribosomes per unit volume of cytoplasm, the fewest mitochondria, the smallest Golgi bodies and relatively thick walls [2, 3, 5, 12, 17, 21, 24]. Nevertheless many kinds of experimental treatments will bring them to a state where they will divide at rates comparable to those in the rest of the meristem. These experimental treatments always involve some impairment of the reproductive future of the dividing cells. Some of them are such as occur in nature - the lowering of temperature to 5°C, for example.

Another feature brought to light by these studies is that the part of the meristem responsible for producing all the cells of the root cap (the cap initials) is especially sensitive to interference with the normal cell cycle. This is true in Zea where the cap initials have a much higher rate of mitosis than other cells and the G_1 phase of the mitotic cycle is largely eliminated by initiating DNA synthesis in telophase[8, 11]. It is not known if the cap initials are hypersensitive in other root meristems where they do not divide especially rapidly.

In the root tip of Zea there are cells that divide at what we may call a normal rate, cells that divide fast and cells that divide slowly or not at all. There are regions where few cells divide and regions where all cells divide. This pattern of proliferation is related to the anatomy of the root apex. What is it that determines rates of mitosis to produce this complex design? This paper describes some of the facts about the meristem and the experiments done to understand the control of cell proliferation in a multicellular system.

MATERIALS AND METHODS

The primary roots of Zea mays cv. Golden Bantam grown in the dark and of standard length have been used for most of this work and four regions of the meristem have been investigated. These are (1) the cap initials (CI) which constitute a disc of about 7000 cells at the proximal margin of the root cap; (2) the quiescent centre (QC), a hemispherical group of about 600 cells at the poles of the stele and cortex; (3) the cells of the stele just above the quiescent centre (ST1) and (4) the cells of the stele 200 µm above the cap boundary (ST2), well within the meristem which extends a further 1000 µm towards the base. In the Figures only the distal region of the stele, ST1, is represented to avoid confusion.

The first step was to discover the rates of mitosis in the various parts of the meristem. This has been done in two ways, by accumulating colchicine metaphases and by labelling mitotic nuclei.

The colchicine technique depends on the fact that the drug stops cell cycles at metaphase for a while, but otherwise has no effect on mitotic cycles up to metaphase. Measurements of the rate of accumulation of metaphases over a short time such as three hours enable us to calculate the durations of the mitotic cycle making some assumptions about the type of proliferation. The parameter found is the average duration for the region scored, equivalent to the cell-doubling time because metaphases are counted as the fraction of all cells. It can be used to find average rates of mitosis in dividing cells when the proportion of cells dividing (the growth fraction) is known.

In the labelled mitosis technique the roots are given ^{3}H-thymidine for a 30 minute pulse and then grown in a medium free of radioactivity and sampled at intervals. The duration of the mitotic cycle is determined from the interval between the ascending limbs of the damped sinusoidal curve of the number of labelled mitoses expressed as a percentage of all mitoses. It applies only to those cells which have relatively short cycles. It disregards those cells that are not in a mitotic cycle and those with cycles much slower than the average for the region. The pulse-labelled mitosis technique therefore gives a rate of mitosis different from the metaphase accumulation figures. This fact becomes especially important in the quiescent centre where a large proportion of the cells are out of cycle. Pulse-labelled mitoses enable us also to measure the durations of the four phases of the mitotic cycle, G_1, S, G_2 and M, in the fast cycling cells of each population. The measurements have to be made over two mitotic cycles and so the method is not available when rates of mitosis are changing rapidly.

By counting various other cells in the pulse-labelled roots we may find out what proportion of the cells are cycling and what proportion have the short cycle displayed by the labelled mitosis curve. This enables us also to determine the average rates of mitosis or cell-doubling times which are comparable with the results from accumulating metaphases. I have described the calculations involved in detail elsewhere[15].

In the X-ray experiments reported here, the roots have been given a heavy, but non-lethal, dose of 1700 or 1800 rads delivered at 130 or 110 rads per minute respectively from a machine operating at 75 or 70 kV. Zea, with only 5pg of DNA per diploid nucleus, is less sensitive to ionizing radiations than many other plants and 1800 rads is equivalent in effect to about 360 rads given to the roots of Vicia faba, for example.

In some experiments the DNA content of nuclei has been compared by using a Barr and Stroud integrating microdensitometer on Feulgen-stained preparations. Where possible this has been done on squashes, but sections have to be used when the position of the nucleus is critical.

RESULTS

1. Normal Cell Cycles

The rates of mitosis at 23°C for the four regions are presented in Table 1, both the average for all the cells (the cell-doubling rate) and also the value for the fast cycling cells displayed by the labelled mitosis curve. The cap initials produce some fifteen times the number of cells per cell as the quiescent centre and the stele some 10 to 14 times. G_1 accounts for most of the variation in the cell cycles, constituting over half the cycle in the quiescent centre and being eliminated in the cap initials by pushing back the onset of DNA synthesis into the previous mitosis. This is indicated by a negative value of -0.4 hours for G_1 in this experiment and accounts for the presence of labelled telophases in some of the cap initials after brief exposures to ^{3}H-thymidine[11].

TABLE 1

DATA FROM THE PULSE-LABELLING OF ROOTS OF ZEA MAYS WITH ^{3}H-THYMIDINE AT 23°C IN WATER.

	Cl	QC	ST1	ST2
Cells formed per cell per day	1.5	0.1	1.0	1.4
Cells formed per fast cycling cell per day	2.3	0.6	1.7	1.7
Percentage of cycle spent in G_1 in fast cells	0	54	19	25
Percentage of cells that divide	86	44	82	99
Percentage of cells with fast cycle	67	17	58	82

Although the cap initials have such a high rate of mitosis, 14% of them do not contribute to cell proliferation at all and about 20 per cent of the cycling cells have slow cycles that do not enter into the determination of the "fast" cycle duration. Less than half the quiescent centre cells divide and 60 per cent of those that do divide have very long cycles. In the stele the number of cycling cells approaches 100 per cent at 200µm from the cap boundary.

2. X-irradiation

The differences in mitotic cycles in the various regions of the meristem lead to differences in behaviour when the root is subjected to stress. The first indication of this came after X-raying roots. Normally roots fed for long periods show no labelling of the nuclei of most of the cells of the quiescent centre whereas the nuclei of the surrounding cells are mostly labelled. After a heavy dose of X-rays there is a reversal in the autoradiographic pattern after feeding the plants with precursors of DNA (Clowes, 1959). Many normally meristematic cells stop synthesising DNA and therefore show no labelling after feeding for periods of one day or more whereas the cells of the quiescent centre start synthesising DNA and become labelled. Such a change in the autoradiographic pattern begins to be obvious at four days after doses of about 1000 rads in Zea and, by six or eight days, the cells of the quiescent centre may be almost the only cells with nuclear labelling. This explains several observations that were puzzling before the discovery of the quiescent centre, for the result of the change shown by the labelling pattern is that a new meristem is created from the formerly quiescent cells.

The phenomenon noted above may also be followed by measuring rates of mitosis in the various regions at intervals after irradiation by the colchicine technique[7, 16]. The average rate in the quiescent centre goes up from 0.14 divisions per day to 0.39 after receiving 1800 rads. In the cap initials, on the other hand, the rate drops from 2.00 to 0.18 divisions per day and, in the stele just above the quiescent centre, from 0.86 to 0.17. By nine days after irradiation the new meristem has established itself in many roots with a new quiescent centre of its own.

The changes in labelling pattern and rates of mitosis suggest that the quiescent centre must somehow be protected against the radiation and this is confirmed by assessing the damage done to the nuclei in the various regions. Figure 1 shows this after four

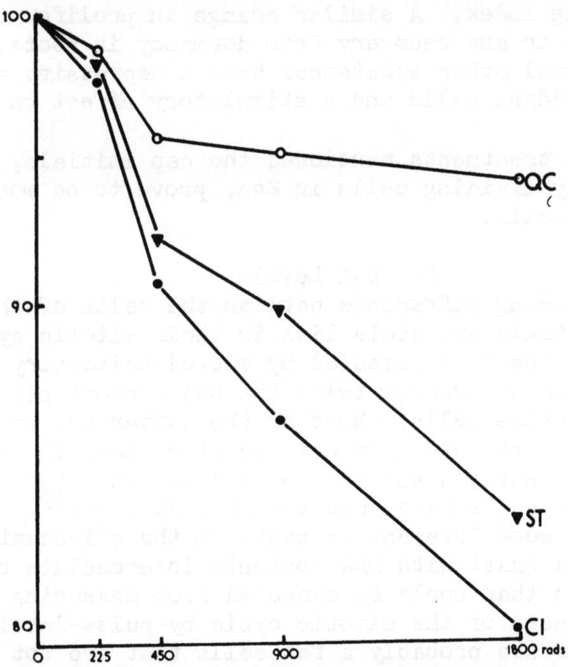

FIGURE 1. FRACTION OF CELLS WITH NO MICRONUCLEI AFTER FOUR DOSES OF X-IRRADIATION. (AREAS UNDER CURVES [SIX DAYS] AS A PERCENTAGE OF UNIRRADIATED CONTROLS).

different doses of X-rays for micronuclei, which result from mitosis in a cell with a broken chromosome and therefore form a guage of nuclear injury. It might be thought that anoxia could provide protection to the quiescent centre, but this is not so because the oxygen enhancement factor for lethal damage to roots is the same as that for reduction in root growth caused by lower doses. Cells that were anoxic could not be further protected by making the whole meristem anoxic[20]. The resistance is not inherent in the cells of the quiescent centre, for their descendants, which enable the root to recover from severe irradiation, are as sensitive as other meristematic cells.

3. Other Treatments

Similar differences in sensitivity among the regions of the meristem are found when roots are given quite different treatments. Transferring seedlings of Zea from 20°C to 5°C for seven days causes the quiescent centre to be stimulated into mitosis when conditions ameliorate whereas other regions suffer a depression in their rates of mitosis. The change in temperature, like X-rays, produces chromosomal damage and this is most marked in the rapidly dividing cells where they acquire a lower than normal labelling index (percentage of cells labelled) after feeding with ^{3}H-thymidine whereas the quiescent centre cells acquire a higher than normal labelling index. A similar change in proliferative roles probably occurs in any recovery from dormancy in roots. 5-aminouracil and several other substances have a depressive effect on the rapidly dividing cells and a stimulatory effect on the quiescent centre.

In all the treatments mentioned the cap initials, which are the most rapidly dividing cells in Zea, prove to be more sensitive than any other cells.

4. DNA Levels

The outstanding difference between the cells of the quiescent centre, cap initials and stele lies in their mitotic cycles. This is reflected in the fact revealed by microdensitometry that the cap initials have an average twice the DNA content per nucleus of the quiescent centre cells. Most of the latter are at the 2C level and most of the cap initials are at or near the 4C level. Some of the cap initials exceed the 4C value and some of the quiescent centre cells have less than the 2C value[12].

A point of some interest is that, in the quiescent centre of Zea, more nuclei exist with DNA contents intermediate between the 2C and 4C levels than could be expected from measuring the durations of the phases of the mitotic cycle by pulse-labelling. This means that there are probably a few cells that are not in the S phase of the mitotic cycle with DNA contents intermediate between 2C and 4C. We know from autoradiography that the quiescent centre develops gradually during the development of the root and some cell cycles may therefore have been stopped in S by whatever induces quiescence in the embryo or lateral root primordium. The cessation of cycling must be brought about in a different way here compared with that in differentiating cells at the proximal margin of the meristem because the latter stop their DNA cycles at the

4C level and may even proceed to higher DNA levels.

The high DNA level of the nuclei of the cap initials in Zea is reduced by the last mitosis and the differentiating cells of the cap possess only the 2C level[12]. It is not until they nearly reach the periphery of the cap that the average DNA content attains the 4C level again with some cells remaining at 2C and others reaching a maximum of 8C by the time the cells are about to be sloughed off.

The information we already have on mitotic cycles and DNA content of cells suggests that they are responsible for the differences in sensitivity to X-rays and other treatments within the meristem. These treatments all result in a change of nuclear damage which could impair the reproductive integrity of cells. A cell with the 2C level of DNA clearly presents a smaller target to ionizing radiation than diploid cells in S, G_2 or M and similar differences in sensitivity between phases of the mitotic cycle probably occur among cells subjected to other treatments. So far as regeneration of the root is concerned, the differential response of the regions of the meristem is amplified by the necessity for several contiguous cells to retain proliferative integrity.

5. Immediate response to X-rays

In addition to the long-term reactions to X-rays described above there are some reactions which occur immediately and which throw some light on the control of cell division. In Zea the mitotic index in the quiescent centre rises immediately after irradiation and reverts to near normal within two hours[13]. In one set of experiments the mitotic index rose from 3.3 to 7.3 per cent 15 minutes after the end of a 1800 rad dose which took 16 minutes to deliver. It then climbed to 13.4 per cent at 30 minutes and then fell steadily. The cap initials and stele showed a steady fall throughout the first two hours from 13.9 to 2.3 and from 14.8 to 4.9 per cent respectively. In another set of roots grown in a different way (in sphagnum and irradiated in situ) the mitotic index of the quiescent centre rose to a maximum immediately at the end of the irradiation (1700 rads which took 13 minutes to deliver). It then fell to a minimum at six hours then rose again during the rest of the day (Figure 2). In this experiment the cap initials also showed an immediate rise followed by a dip at three hours and then a steady rise at the end of the first day. Neither part of stele showed any initial rise, but there was a rise later in the day after the early fall.

The stimulation of cycling in the quiescent centre is also shown by a rise in labelling index from 7.5 to 23.4 per cent in two hours after irradiation whereas the labelling index fails to rise in the other regions. Labelling also shows that some cells proceed from S to metaphase within 30 minutes in both the quiescent centre and cap initials. But the most interesting discovery is that some of the cells of the quiescent centre can enter mitosis within 46 minutes from G_1 without synthesising DNA at all. This

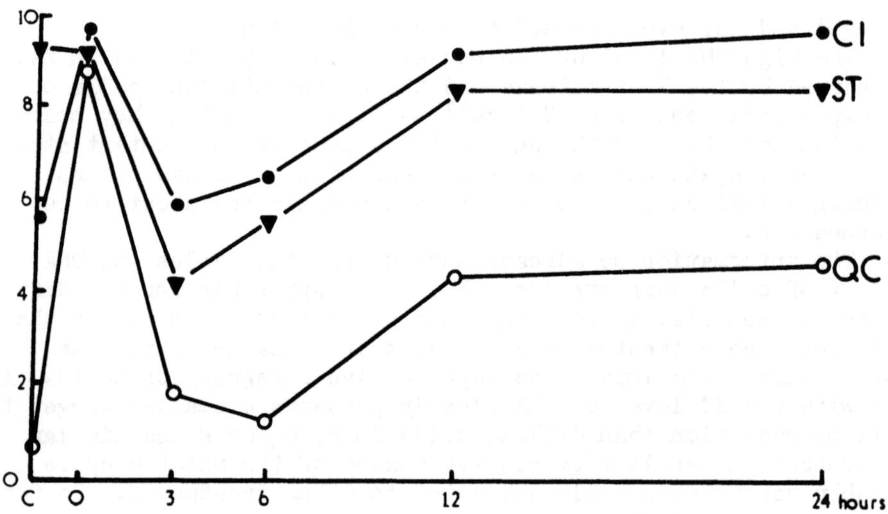

FIGURE 2. PERCENTAGE OF CELLS IN MITOSIS AT INTERVALS AFTER X-IRRADIATION (C, unirradiated controls).

is confirmed by finding prophases and metaphases with less than the 4C level of DNA. There is also some reason to believe that this may occur in normal unirradiated roots as a rarer phenomenon.

It is not possible to measure rates of mitosis immediately after irradiation because the measurement itself by metaphase accumulation takes a minimum of three hours. Results obtained at intervals during the first day after irradiation are given in Figure 3. In the quiescent centre there is a steady rise in the rate to some six times its normal value. In the other regions of the meristem, on the other hand, there is a quick fall to a very low value followed by some recovery.

6. Control of mitosis by root caps

Another experiment which has implications for the study of cell proliferation involves removing part or all of the root cap. In Zea the cap may be taken off more or less intact without removing any other cells. Its meristem at the proximal margin is a plane of weakness and the thick-walled cap boundary protects the rest of the root when the cap is prized off. This fact has been exploited by people interested in geotropism because the cap appears to control the geotropic response of the root, but its removal does not affect the growth rate of the root [19, 22].

The result of removing the whole cap is to stimulate mitosis in the quiescent centre from 0.12 divisions a day to 0.46 (measured by metaphase accumulation) at five hours after the operation at 23°C. In the stele the rates fall slightly. The quiescent centre regenerates a new set of cap initials and eventually this produces a normal cap and the quiescent centre reverts to mitotic quiescence.

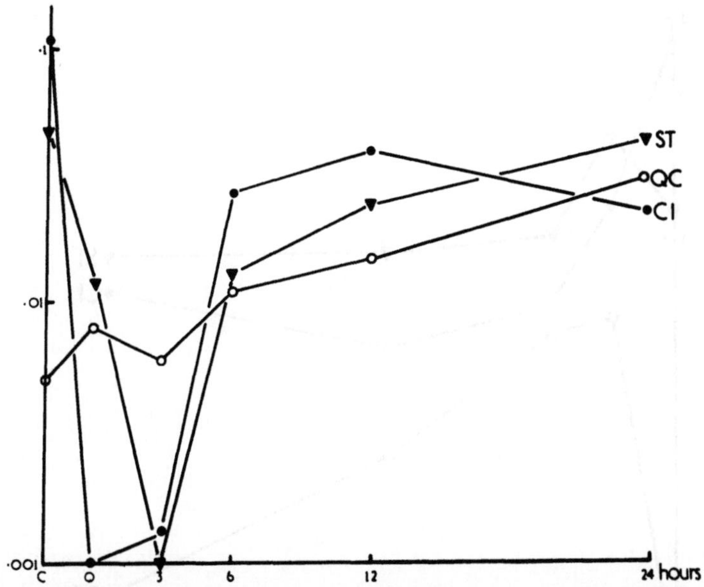

FIGURE 3. NUMBER OF MITOSES PER HOUR AFTER X-IRRADIATION. The times given are the intervals between irradiation and the start of metaphase accumulation; C, unirradiated controls.

When the distal half of the cap is cut off with a knife the change in cell division is even more striking (Figure 4). Here the cap initials remain intact, but nevertheless ther proliferation tails off and they are replaced by cells derived from the renewed mitosis in the quiescent centre. The new cells burst through the cap boundary, form a new disc of cap initials and take over the activity of the old initials which are pushed towards the tip and differentiate. This occurs within two to three days at 23°C in many individuals and leads to a rise in the rate of mitosis in the region.

DISCUSSION

The first impression that many of these experiments give is that there is competition between the regions of the meristem for some substance essential for cell growth or division and that, in normal circumstances, the quiescent centre is held non-meristematic with most of its cells in G_1 by starvation or some similar phenomenon. As soon as this is relived, the cells of the quiescent centre revert to a proliferative phase. A view based on this impression is that the quiescent centre is at the end of a supply route from the shoot and consequently is starved. The high rate of mitosis and growth in the cap, whose 10,000 cells are completely renewed every 24 hours at 23°C, could be explained as due to the

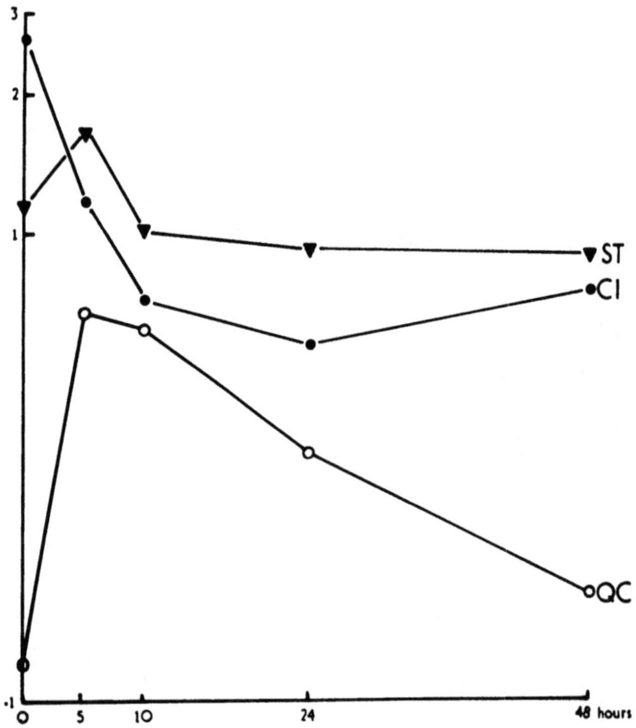

FIGURE 4. NUMBER OF MITOSES PER DAY AT INTERVALS AFTER REMOVING THE DISTAL HALF OF THE ROOT CAP.

recycling of substances made available from the moribund cells on the periphery of the cap. This view is supported by three of Barlow's observations. The duration of the S period increases steadily in the stele as the tip is approached. Removing half the endosperm slows the growth rate of seedling roots. The grain count over nuclei in S is higher in the quiescent centre than further up the root and so the externally supplied labelled precursor is presumably diluted by internal supplies less in the quiescent centre than further up the root. Barlow[1] used ^{3}H-thymidine and the experiment ought to be reexamined in the light of MacLeod's[23] work on thymidine Kinase induction.

When thymidine is supplied to the endosperm of seedlings, the grain count over nuclei also falls off towards the tip, but the cap initials are labelled as much as the distal dividing cells of the stele. Also it is possible to supply the quiescent centre and stele with DNA precursors from the distal half of the cap[14]. Thus, although the quiescent centre may be at the end of supply lines from the shoot and cap, this does not appear to be the cause of quiescence.

Somehow the demand for supplies is regulated within the meristem itself. This does not prevent a nutritional stimulus from causing renewed meristematic activity in the quiescent centre when injury to the meristematic cells relaxes the demand for nutrients or hormones elsewhere. Van't Hof[26, 28] has shown how a spurt of carbohydrate can push cells into S in root tips held artificially in a non-cycling condition. The effect of cutting off the distal half of the cap is especially interesting in this connexion because no dividing cells are lost in the process, yet it has a rapid and complex effect on the rates of mitosis.

It is possible that pressure exerted from all sides by the growing cells may restrict growth, and therefore mitosis, in the quiescent centre. This view would account for the different shapes of quiescent centres in roots with different cell patterns. Interference with cell proliferation could relieve the pressure and allow the quiescent centre to expand and its cells to divide. Some preliminary studies have also shown that when root caps are removed and pressure is applied artificially to the end of the root the quiescent centre may remain quiescent instead of becoming active as it would do without pressure.

I have omitted much consideration of the role of hormones in the control of cell proliferation patterns because I find it difficult to relate what information we have to a multicellular system as complex as a root meristem. The work of Torrey[25] and Van't Hof[27] among others shows how important growth substances must be, but we need more refined assay methods before we can use this knowledge in investigating natural control systems such as must exist in meristems.

Whatever it is that controls cell proliferation in root meristems exerts a finely balanced equilibrium on the different regions. The equilibrium can be upset to produce a visible change in cell cycles within a few minutes and whatever upsets the equilibrium can force the cells out of their normal mitotic cycles of DNA synthesis and mitosis.

CONCLUSIONS

The meristem of a root has been divided into regions so that the organization of cell proliferation may be analysed. Rates of mitosis and the proportion of dividing cells have been determined. Rapidly dividing cells and slowly dividing cells exist side by side and therefore the system is useful for investigating the control of cell proliferation. This has been done by subjecting the meristem to various stresses which upset the fine balance between cells. The impression given by these experiments is that differences in proliferation are controlled from within the meristem rather from outside.

REFERENCES

1. Barlow, P.W. (1969) Organization in root meristems. Oxford University, D.Phil. thesis.
2. Barlow, P. (1969) Nucleolar RNA content of cells in the root apex of Zea mays. Chromosomes Today ed. Darlington &

Lewis 2:121-125.
3. Barlow, P.W. (1970) RNA synthesis in the root apex of Zea mays. J. exp. Bot. 21:292-299.
4. Clowes, F.A.L. (1954) The promeristem and the minimal constructional centre in grass root apices. New Phytol. 53:108-116.
5. Clowes, F.A.L. (1958) Protein synthesis in root meristems. J. exp. Bot. 9:229-238.
6. Clowes, F.A.L. (1959) Reorganization of root apices after irradiation. Ann. Bot. 23:205-210.
7. Clowes, F.A.L. (1963) X-irradiation of root meristems. Ann. Bot. 27:343-352.
8. Clowes, F.A.L. (1965) The duration of the G_1 phase of the mitotic cycle and its relation to radiosensitivity. New Phytol. 64:355-359.
9. Clowes, F.A.L. (1965) Synchronization in a meristem by 5-amino uracil. J. exp. Bot. 16:581-586.
10. Clowes, F.A.L. (1967) The quiescent centre. Phytomorphology 17:132-140.
11. Clowes, F.A.L. (1967) Synthesis of DNA during mitosis. J. exp. Bot. 18:740-745.
12. Clowes, F.A.L. (1968) The DNA content of the cells of the quiescent centre and root cap of Zea mays. New Phytol. 67:631-639.
13. Clowes, F.A.L. (1970) The immediate response of the quiescent centre to X-rays. New Phytol. 69:1-18.
14. Clowes, F.A.L. (1970) Nutrition and the quiescent centre of root meristems. Planta (Berl) 90:340-348.
15. Clowes, F.A.L. (1971) The proportion of cells that divide in root meristems of Zea mays. Ann. Bot. 35:249-261.
16. Clowes, F.A.L. & Hall, E.J. (1962) The quiescent centre in root meristems of Vicia faba and its behaviour after acute X-irradiation and chronic gamma irradiation. Radiat. Bot. 3:45-53.
17. Clowes, F.A.L. & Juniper, B.E. (1964) The fine structure of the quiescent centre and neighbouring tissues in root meristems. J. exp. Bot. 15:622-630.
18. Clowes, F.A.L. & Stewart, H.E. (1967) Recovery from dormancy in roots. New Phytol. 66:115-123.
19. Gibbons, G.C.B. & Wilkins, M.B. (1970) Growth inhibitor production by root caps in relation to geotropic responses. Nature 226:558-559.
20. Hall, E.J., Lajtha, L.G. & Clowes, F.A.L. (1962) The role of the quiescent centre in the recovery of Vicia faba roots from irradiation. Radiat. Bot. 2:189-194.
21. Juniper, B.E. & Clowes, F.A.L. (1965) Cytoplasmic organelles and cell growth in root caps. Nature 208:864-865.
22. Juniper, B.E., Groves, S., Landau-Schachar, B. & Audus, L.J. (1966) Root cap and the perception of gravity. Nature 209:93-94.

23. MacLeod, R.D. (1971) Thymidine Kinase activity and its relationship to DNA synthesis in the primary root of Vicia faba. Ann. Bot. 35:237-248.
24. Pilet, P.E. & Lance-Nourarède, A. (1965) Quelques characteristiques structurales et physiologiques du méristème radiculaire du Lens culinaris. Bull. Soc. Franc. de Phys. Veg. 11:187-201.
25. Torrey, J.G. (1963) Cellular patterns in developing roots. Symp. Soc. exp. Biol. 17:285-314.
26. Van't Hof, J. (1968) Control of cell progression through the mitotic cycle by carbohydrate provision I. Regulation of cell division in excised plant tissue. J. cell. Biol. 37:773-778.
27. Van't Hof, J. (1968) The action of IAA and kinetin on the mitotic cycle of proliferative and stationary phase excised root meristems. Expl. Cell Res. 51:167-176.
28. Van't Hof, J. (1970) Metabolism and the prolonged retention of cells in the G_1 and S period of the mitotic cycle of cultured pea roots. Expl. Cell Res. 61:173-182.

-DISCUSSION-

D'AMATO - After irradiation some of the cells of your root system are G_1 (2C) cells which enter mitosis. This means essentially that the chromosomes are one-chromatid as they enter mitosis. Have you cytological evidence to support this?

CLOWES - No, I do not, except that there are telophase nuclei and anaphases with 1C amounts of DNA.

SAVAGE - The mitotic index values for the various regions of the root are expressed relative to one another. With radiation changing cycle time what criteria do you use for delineating the quiescent center?

CLOWES - From experience, I know the region of the root which the quiescent center occupies. I can, therefore, determine the mitotic activity of the cells in those regions. I have assumed no change in size of the quiescent center.

VAN'T HOF - Are the 1C cells perpetuated? One would think that if they were important in repopulating the meristem their progeny would be similar in content but greater in number.

CLOWES - If by chance the DNA content per cell drops to 1C then there are mechanisms for pushing the DNA content back up as Dr. D'Amato's work has shown. A similar phenomenon is observed in the tapetum of stamens.

VAN'T HOF - Do you observe a variation in the duration of S among the various cell populations you study?

CLOWES - No. The conspicuous difference is in G_1.

TORREY - You reported that the immediate derivatives of the root cap initials have DNA contents greater than 4C. Do you observe division figures with greater than 4C amounts?

CLOWES - This is a difficult question to answer. One has to work from sections to know the exact location of cells and it is not possible to determine DNA contents of mitotic figures accurately enough from sections. I might add that when the derivatives cease dividing their DNA content is 2C and they remain so until they arrive at the root cap periphery where their average DNA content is observed to be 4C.

CONGER, A. - Is the DNA content per cell of the outer layer of the root characteristically different from that of the meristematic region of the root?

CLOWES - At comparable levels of the root the DNA values are similar because the epidermal cells do continue to divide; but within the meristem one will get considerable variation at a particular level depending upon what tissue you are looking at.

JUNIPER - Maize unfortunately does not have chromosomes which lend themselves to standard cytological analyses; but, it is possible through the use of the high voltage electron microscope and thick sections (about 4 microns) to create images of whole chromosomes including centromeres and chromatids.

BALL - Is there any significance to the presence of micronuclei in irradiated cells of the quiescent center? Do these cells with micronuclei die?

CLOWES - Radiation stimulates the cells of the quiescent center to divide. It also induces chromosome aberrations which can result in the presence of micronuclei.

D'AMATO - There is a considerable amount of evidence that cells with chromosome aberrations can continue to undergo mitosis; for example in *Triticum*.

CONGER, A. - The conventional answer to Dr. Ball's question is that the cell with the micronucleus has not really suffered any genetic loss -- all of the chromosome pieces are in one cell and the cell therefore contains all its genes.

SAVAGE - Grote and Revell (IVth International Congress of Radiation Research, Evian, 1970. Abstract 332) have been able to study *live* mammalian single cell cultures for the presence of chromosomal

SAVAGE - aberrations; there is a 1:1 correlation between the presence of micronuceli and the induction of lethality and the same is true for irradiated *Tradescantia* microspores.

DAVIDSON
D'AMATO - (in unison) But not with other plant material.

DAVIDSON - With root tips it is possible to get whole clones of cells which have deficient chromosomes.

VAN'T HOF - This raises an interesting point about single cells in culture and cells contained in a multicellular structure. Perhaps the demand on the genetic machinery of single cells is much greater and they cannot tolerate even slight deletions. The cells of a tissue can depend to some extent upon their neighbors for support.

MORPHOGENETIC ASPECTS OF THE DEVELOPMENT OF MERISTEMS IN SEED EMBRYOS

F. D'Amato

Institute of Genetics, The University, Pisa, Italy

INTRODUCTION

Evolution has favored the establishment, in Phanerogams, of a propagation and survival organ, the seed; this contains, in a resting state, the embryo, a young plant which resumes growth and differentiation at seed germination. The embryo shows a variety of developmental patterns and may attain, at seed maturity, different sizes and degrees of differentiations. The fully developed embryo commonly consists of an axis bearing one or more cotyledons and, at its opposite ends, the primordium of the root and the primordium of the shoot. The root and shoot primordia vary in degree of differentiation from simple apical meristems to embryonic organs: the embryonic root -or radicle- and the embryonic shoot -or plumule- with one or more leaf primordia. In the grass family, the highly differentiated embryos also bears the primordia of adventitious roots.

The initiation and development of meristems in an embryo raises a variety of questions, for some of which I will try to offer an answer. These are the following:

(1) In developing embryos, cell proliferation in the meristems can only proceed until a stage of maturation when the seed dries and becomes quiescent. An immediate question to be answered is: in what stage of the nuclear cycle can the meristematic cells of

an embryo be stopped by the process of seed maturation?

(2) Can techniques so efficiently applied to a dynamic analysis of cell populations in root apices -*e.g.* nuclear labeling with radioactive thymidine and DNA cytophotometry- help us in understanding the primary event(s) responsible for the block of meristem cell proliferation in maturing embryos? Information on this aspect is basic to investigations which may attempt at closely analyzing the temporal sequence of initiation and development of the several meristems during embryogenesis.

(3) Can irradiation of resting (=dry) seeds provide further information useful to a better characterization of some nuclear cycle phases present in cells of embryonic meristems?

NUCLEAR CONDITIONS IN MERISTEMS OF RESTING SEEDS

It has been generally assumed that meristems in resting seeds are composed of cells with nuclei in the pre-DNA synthesis phase (G_1). This assumption has been favored by the observation that, in some species, irradiation of resting seeds induces predominantly or exclusively chromosome-type aberrations, which may be seen in the first mitoses in the root tips when the seeds are germinated.[14,19,24] However, in 1963, we[3] found that X-irradiation of dry seeds (caryopses) od *durum* wheat (*Triticum durum*) resulted in a rather high frequency of chromatid-type aberrations in the first mitotic cycle post-germination. Our suspicion that G_2 (post-DNA synthesis phase) cells might be present in the embryonic radicles of *durum* wheat was confirmed by two sets of data. First, cytophotometric determination of nuclear DNA content revealed the contemporary occurrence of 2C and 4C nuclei in all of the 10 resting radicle meristems analyzed. In some of the radicle meristems, we also measured a few nuclei with 2C-4C intermediate values. Secondly, germination in ^{3}H-thymidine of irradiated resting seeds showed that practically all (58/61) mitoses with chromosome aberrations were labeled, whilst of the mitoses with chromatid aberrations part were unlabeled (82/142) and the rest were labeled. This data, in providing the first information on the radiation response of G_1 nuclei in resting seeds, also demonstrated that most, if not all, 2C and 4C nuclei in resting radicle meristems were actually nuclei in the G_1 and G_2 phases of the DNA cycle.

Since 1963, we have continued our work on embryonic meristems and directed first of all attention to a species of great radiobiological interest, Vicia faba.[9] Using X-irradiation of resting seeds combined with DNA cytophotometry and nuclear labeling with ^{3}H-thymidine in two varieties of Vicia faba, we demonstrated that the relative proportion of 2C (G_1) and 4C (G_2) cells in the radicles meristem of resting seeds showed great variation from seed to seed within a variety and the average percentage of 4C (G_2) cells in the two varieties was different. We also found that the percentage of unlabeled mitoses in germinating seeds (G_2 cells in the resting embryo) was very high at 22 hours of seed soaking and progressively decreased to nearly 0 at 54 hours (fig. 3).

Further work in our Laboratories has shown that, in other species, the radicle meristem in resting seeds is composed of 2C (G_1) cells exclusively (Pinus pinea, Lactuca sativa, Helianthus annuus[10,11]) or almost exclusively (Allium cepa[8]). 2C cells only are also found in the embryonic shoot apex of Pinus and in the cotyledons of Pinus and Lactuca embryos.[10]

DNA labeling and/or DNA cytophotometry, sometimes combined with irradiation, have been used by other workers on dry seeds: they have provided informations on the nuclear conditions in meristems of resting embryos in other plant species: Zea mays,[25] Haplopappus gracilis[23], Triticum aestivum[28], Pisum sativum.[5,6] In this last species, the embryonic root apex was found to contain not only 2C (G_1) and 4C (G_2) nuclei but also nuclei with 2C-4C intermediate values (S phase); the frequency of S nuclei was much different in the two harvest years investigated (5% and 41% in the seed crops 1963 and 1965 respectively).

SEQUENCE OF DNA SYNTHESIS AND MITOSIS IN DEVELOPING EMBRYONIC MERISTEMS

The nuclear conditions attained by cells in the embryonic meristems at seed maturation are the outcome of the sequence of DNA synthesis and mitosis in the last stages of the maturation process.

In 1967, Brunori[7] studied the development of the primary root meristem in the embryo of Vicia faba. Samples of developing seeds were collected at weekly intervals during the forty-day period preceding seed harvest (ripe seeds). From each seed sample, Feulgen squashes of primary root meristems were prepared for analysis of mitotic index and cytophotometric determination of DNA content in interphase nuclei. At each collection time, the water content of the cotyledons was determined, in view of the role that progressive drying plays in the ripening of seeds. The most significant observations made by Brunori[7] are reported in

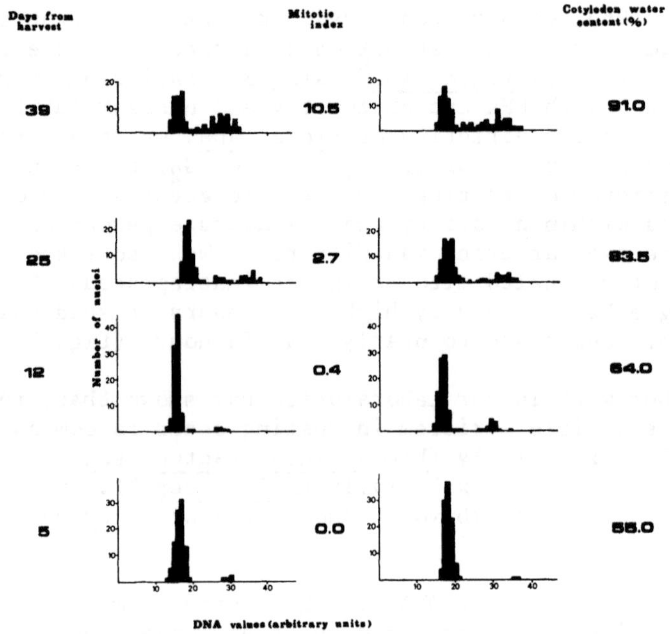

Fig. 1. Distribution of DNA contents in interphase nuclei of root meristems of embryos of Vicia faba at different times from seed harvest (based on data from Ref. 7).

fig. 1. At 39 days from seed harvest, the mitotic index in the root meristems was 10.5 -quite comparable to the mitotic index which can be found in root tips of Vicia faba seedlings- and the interphase nuclei were comprised in the three expected classes of 2C (G_1), 2C -4C (S) and 4C (G_2). At 25 days from seed harvest, the mitotic index dropped to 2.7 and the frequency of S nuclei was clearly reduced, indicating some depression in the DNA synthetic activity. This was followed by inhibition of DNA synthesis, as manifested by the two discrete DNA peaks - 2C (G_1) and 4C (G_2) - obtained with the DNA cytophotometry of the root meristems collected at 19 and 12 days from seed harvest (only the graphs of this second collection are reported in fig. 1). At these same times, there was a residual mitotic activity (mitotic index: 0.6 and 0.4 respectively) which reduced the proportion of 4C (G_2) relative to 2C (G_1) cells.[7]

Thus, in the primary root meristem of developing embryos of Vicia faba, DNA synthesis is stopped earlier than mitosis. If water content is taken as a parameter, DNA synthesis is clearly more sensitive to decreased water supply than mitosis: the two processes are stopped respectively at 75% and 55% cotyledon water content.[7] Since, at seed maturity, primary root meristems of Vicia faba embryos can be left as a mixed population of 2C (G_1) and 4C (G_2) cells,[9] it may be assumed that the time interval between inhibition of DNA synthesis and inhibition of mitosis is not long enough to allow the entry into mitosis of all G_2 cells; in other words, the "phase of depletion of G_2 cells"[2,10] is incomplete. It seems plausible that in other species showing 2C (G_1) and 4C (G_2) cells in the embryonic root apices of dry seeds, the sequence of DNA synthesis and mitosis during embryo development and maturation may follow a pattern similar to that of Vicia faba. In species of this group, the depletion phase would be a rather critical phase liable to environmental influences: the great variability in relative frequency of 2C and 4C cells in the materials investigated so far would support this view.

Species in which the embryonic meristems in resting seeds are exclusively composed of 2C cells apparently belong to a distinct category: during embryo development and maturation conditions must prevail which allow a 100% efficiency of the depletion phase. As discussed previously,[2,10] the situation of an uniform 2C meristem cell population merely reflects a strict control of the sequence of DNA synthesis and mitosis, which is probably bound to the genetic make-up of species or taxonomic groups.[15,16] Several mechanisms might be invoked for a 100% efficiency of the depletion phase. Among these, two seem to be operative: first, completion of the development of a meristem at a time in embryogenesis, when some factors has yet to become limiting and second, a change in the duration of the S phase relative to the duration of mitosis.

In this connection, our work on the development of meristems in the embryo of durum wheat[2] is instructive. Seeds in the final stages of development were collected from durum wheat plants. At each collection time, part of the seeds were used for determination of fresh and dry weights and water content; from other seeds the embryos were excised, treated with ^{3}H-thymidine for one hour and fixed in acetic alcohol. Feulgen-stained histological sections and squashes were used for studying separately the following meristems: apex of primary root; apex of adventitious roots of the first and the second pair; primordium of the first, second and third leaf. Following autoradiography, the slides were scored for labeling index (percent labeled interphase nuclei) and mitotic index (mitotic stages in percent of total cells). At seed maturity, embryos were fixed in acetic alcohol, stained with Feulgen and separate squashes of the same meristems as above and of the shoot apex were prepared. These slides were used for DNA cytophotometry.

It was found that at 20 days before seed maturation (water content, 63%), the shoot apex had already been formed: neither mitosis nor incorporation of thymidine into interphase nuclei occurred. In the other meristems, both DNA synthesis and mitosis were occurring, but their relative rate, as expressed by the labeling index -mitotic index ratio,[12,13] was different. This difference

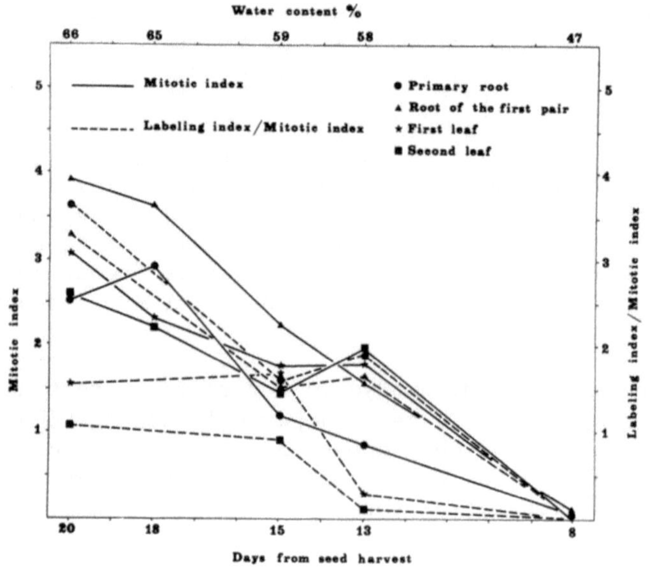

Fig. 2. Mitotic index and labeling index - mitotic index ratio in different meristems of embryos of Triticum durum at different times from seed harvest (based on data from Ref. 2).

became still clearer in subsequent development stages. An analysis of the labeling index -mitotic index ratio revealed two interesting facts (fig. 2). First, root meristems on one side and leaf meristems on the other manifested a contrasting behaviour. Leaving aside the material of the last collection preceding seed maturation, the labeling index -mitotic index ratio was always higher than 1 in the root meristems, but in leaf primordia ratios greater than 1 were found in the first three collections in leaf 1 and in the first collection in leaf 2. The contrast between root meristems and leaf meristems was most striking at 13 days from seed maturity. The very low labeling index -mitotic index ratios in leaf meristems (0.10 and 0.30) clearly showed that DNA synthesis was being stopped earlier than mitosis. This condition, by allowing the entry into mitosis of G_2 cells, favored the depletion of 4C (G_2) cells in the meristems. In fact, DNA cytophotometry in the embryos of mature seeds showed that the three leaf primordia were composed of 2C cells only (Table I). Secondly, in the root meristems - in which the labeling index -mitotic index ratio at late development stages remained greater than 1 - the interval between inhibition of DNA synthesis and inhibition of mitosis was shorter -or much shorter- than in leaf primordia. Consequently, the depletion phase in root meristems was incomplete as demonstrated by DNA cytophotometry in mature seeds (Table I): the depletion phase had occurred with

TABLE I

Frequency of nuclei with 2C and 4C DNA contents in meristems of five embryos in dry (resting) seeds of *Triticum durum*.[2]

Meristems	Embryos No. 1		No. 2		No. 3		No. 4		No. 5	
	2C	4C	2C	4C	2C	4C	2C	4C	2C	4C
Primary root	77	23	86	14	61	39	72	28	80	20
Root of I pair	90	10	91	9	95	5	94	6	92	8
Root of II pair	100	0	100	0	92	8	100	0	100	0
I leaf	100	0	100	0	100	0	100	0	100	0
II leaf	100	0	100	0	100	0	100	0	100	0
III leaf	50	0	50	0	50	0	50	0	50	0
Shoot	30	0	30	0	30	0	30	0	30	0

different efficiency in the three types of root meristems as shown by the relative proportions of 4C (G_2) nuclei in the cell populations.

Thus, in *durum* wheat, different embryonic meristems have a different developmental history, but all tend to be composed, at maturity, of 2C cells predominantly or exclusively. Uniformity of 2C (G_1) cell populations is apparently obtained in two different ways:

(1) Early initiation and development of a meristem, so that its structure is completed before the negative influence of some limiting factor (*e.g.* water supply) may manifest itself: this is the case of the shoot apex, which was found to be the first meristem that completed development during embryogenesis.

(2) A change in duration of the S phase relative to the duration of mitosis, ending in an inhibition of S at an earlier time than inhibition of mitosis: this mechanism operated with full efficiency during the formation of embryonic leaf primordia.

CHARACTERIZATION OF 2C (G_1) CELLS IN RESTING SEED MERISTEMS

We now turn to the third point of the present discussion: the type of induced chromosomal damage in its relation to the nuclear conditions in the resting seed state.

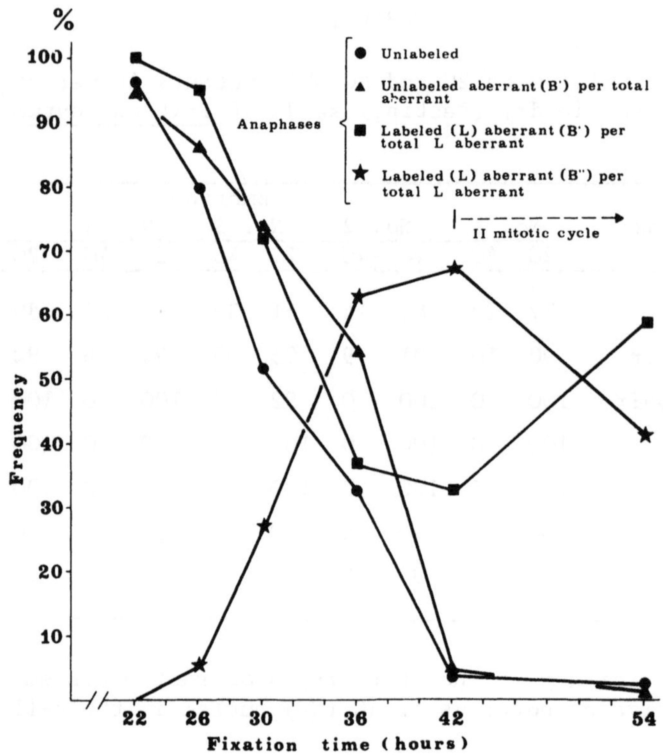

Fig. 3. Response to X-irradiation (1500 R) of G_1 and G_2 cells of the radicle meristem of resting seeds of *Vicia faba* (based on data from Ref. 9).

In our 1963 work on the X-irradiation of dry seeds of *durum* wheat,[3] indication was found that part of the 2C (G_1) cells responded to radiation with chromatid aberrations; the chromatid-type response of G_1 cells in the same material was confirmed by the observations of Avanzi *et al*.[4] on X-irradiated dry seeds; these also showed that the root cells entering mitosis first in the germinating seed were actually cells in G_2 in the resting seed.

Further work was done on dry seeds of *Vicia faba* irradiated with an X-ray exposure -1500 R- which did not alter the sequential release of cells from quiescence to mitosis as demonstrated by germination of the irradiated seeds in ^{3}H-thymidine.[9] It was found that:

(1) at the onset of mitosis in the germinating seeds (22 hours after seed soaking), 96% anaphases were unlabeled, thus representing cells that had completed DNA synthesis (G_2) before the root entered the resting seed state. At later times, the relative proportion of unlabeled to labeled anaphases progressively decreased with onset

of DNA synthesis and mitosis in cells which were in G_1 in the resting seed; but some unlabeled anaphases (2.5%) still occurred at 54 hours (fig. 3).

(2) All unlabeled anaphases bore chromatid aberrations (B'); they constituted the predominating fraction of all aberrant anaphases in the first 36 hours (95.2-54.5%), after which time they decreased sharply and almost disappeared (fig. 3).

(3) Among labeled aberrant anaphases (cells hit by radiation in G_1), some had chromatid breaks (B') and others chromosome breaks (B"). Anaphases with B' greatly predominated in the first 30 hours (100-72%); at later times, anaphases with B" were the most frequent (fig. 3).

These observations led to the conclusion that in Vicia faba, in which the embryonic root apex contains both 2C (G_1) and 4C (G_2) cells, the release of cells from quiescence to mitosis at seed germination occurred according to a well defined temporal progression: G_2 (unlabeled cells, chromatid aberrations), late G_1 (labeled cells, chromatid aberrations), early (or mid) G_1 (labeled cells, chromosome aberrations). Thus, in the resting seeds of Vicia faba[9] and in those of durum wheat[4] as well, cells in the terminal portion of G_1(late G_1) showed the same radiation response (chromatid-type aberrations) as late G_1 in proliferating cells.[20,21,22,29]

The recent availability of seeds having only 2C cells in their seed embryos (Pinus pinea, Lactuca sativa, Helianthus annuus) stimulated further analyses on the X-radiation response of G_1 cells in resting seeds. It was found that, in these three species, G_1 cells responded to radiation predominantly with chromosome break (B") aberrations: in the first mitotic cycle post-irradiation; the percent ratio of B" breaks to total breaks ranged from ca. 94 to ca. 82 in Pinus from ca. 91 to ca. 80 in Helianthus and from ca. 85 to ca. 69 in Lactuca. Although we did not exclude the possibility of the occurrence of late G_1 cells within the G_1 cell population of the seed meristems, we suggested that the B' breaks in Pinus, Helianthus and Lactuca could be, at least in part, the result of an indirect radiation effect.[11] Quite recently, Dubinin et al.[18] have found that in Crepis capillaris, having embryonic root meristems composed of G_1 cells only, gamma-ray irradiation induces chromosome aberrations only in air-dry seeds (moisture content 8.2-3.5%); when artificially dried seeds (2% moisture content and less) are irradiated, a few percent of the induced breaks are of definite chromatid type.

Taken together, the above observations on the radiation response of G_1 cells in resting seeds would show that meristems composed of 2C (G_1) cells only respond to radiation exclusively or predominantly with chromosome aberrations, whilst 2C (G_1) cells

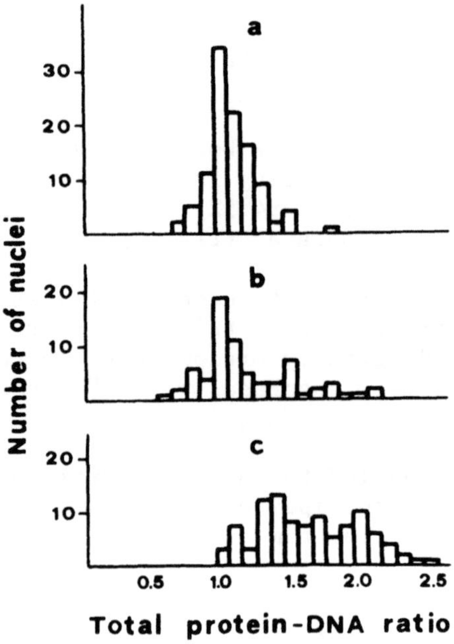

Fig. 4. Distribution of total protein - DNA ratios in 2C nuclei of radicle meristems in resting seeds of (a) Pinus pinea, (b) Triticum dicoccum and (c) Vicia faba (by courtesy of Drs. A. Brunori and G. Martini).

in meristems composed of both 2C and 4C cells respond to radiation with either chromatid (late G_1) or chromosome (early or mid G_1) aberrations. This would mean that meristems made of 2C cells only in the dry seed state are "synchronized in G_1."[18] But, the question arises: is homogeneity in 2C nuclear DNA content a valid criterion for synchronization in G_1?

Since 1970, my Colleagues, Dr. A. Brunori and Dr. G. Martini, are analyzing cytophotometrically histone - DNA and total protein - DNA ratios in root meristem nuclei of resting seeds of several plant species. Nuclei isolated from root apices of dry seeds were first stained with Feulgen and their DNA content was determined cytophotometrically; after mapping of nuclei in photomicrographs, the nuclei were processed for specific staining for either histones[1] or total protein[17] and again measured in the cytophotometer. Considering nuclei with 2C DNA content in any one species, it was seen that the amount of histone relative to that of DNA did not show appreciable variations; the situation was different with total proteins. By permission of Drs. Brunori and Martini, I am showing the data they obtained on three species which represent a typical case each (fig. 4,a-c). In Pinus pinea, in which the embryonic root apex is made exclusively of 2C cells that respond to radiation almost exclusively with chromosome-type aberrations, the distribution of total protein-DNA ratios is rather narrow (fig. 4,a). In Triticum dicoccum, in which the embryonic root apex is made almost exclusively (98-99%) of 2C cells that respond to radiation with either chromatid or chromosome aberrations (Brunori and Martini, unpublished) a rather wide variation in total protein-DNA ratios is found (fig. 4,b). In Vicia faba, in which the embryonic root apex is made of both 2C and 4C cells and the 2C cells respond to radiation with either chromatid or chromosome aberrations, the total protein-DNA ratios within the 2C cell population show a scatter of values (fig. 4,c) comparable to that of Triticum dicoccum (fig. 4,b).

The data in fig. 4, together with the type of G_1 radiation response, would support the assumption that the 2C cells of the embryonic root apex of Pinus pinea are synchronized totally -or almost totally- in early (or mid?) G_1, while those of the embryonic root apex of Triticum dicoccum and Vicia faba are not synchronized, being composed of both early (or mid) and late G_1 cells. The data also stress the role that total protein content of G_1 nuclei might probably have in conditioning the transition from chromosome- to chromatid-type aberration response during G_1.

CONCLUDING REMARKS

The observations we have summarized above clearly show that an asynchronous population of proliferating cells, such as occurs in developing seed meristems, can collect in different phases of the DNA cycle when cell division ceases in conjunction with ripening of the seed. Most commonly, cells accumulate in 2C (G_1): this is shown by the exclusive occurrence of 2C (G_1) cells in the resting seed meristems of some species and by the fact that, in species in which meristems are left as a mixed population of 2C (G_1) and 4C (G_2) cells in the dry seed, 2C cells predominate over 4C. The only exception would seem to be meristems of dry seeds of maize, in which available data indicates that the majority of cells are resting in G_2[25]; but decision on this case must wait for further more thorough analyses. There are also indications that, in exceptional cases, cells of embryonic meristems can be held in S. The very high proportion of S cells in the radicle apex of dry pea seeds of the crop 1965, but not in those of the crop 1963,[6] most probably reflects the action of unfavorable environmental conditions in the last stages of seed maturation. This assumption is supported by our observations on several crops of durum wheat, in which S-class nuclei in resting embryonic radicle apices were only found in some of the seeds of the crop 1962.[3]

Analyses on the onset of DNA synthesis and mitosis at seed germination show that G_1 and G_2 cells present in the dry seed meristems are able to progress in the nuclear cycle and enter mitosis. It is not yet known whether cells held in S in a resting seed meristem can progress normally to mitosis when the seed germinates. This possibility cannot be excluded in view of Van't Hof's experiments on excised pea root tips in culture.[26,27] In stationary phase meristems established in a medium without carbohydrate, cells could be positioned in the S period by transfer in a medium with sucrose and retained in S with hypoxia. Insufficient oxygen for up to 24 hours did not affect the ability of the cells to resume DNA synthesis and enter mitosis when hypoxia terminated. Since in the excised pea root tip system cells collect preferentially in G_1, "the relative probabilities that a cell would remain in a given period would be $G_1 > G_2 >$ (S or M").[26] This situation finds a striking parallel in the nuclear conditions which are established in embryonic meristems when the seed ripens. It remains to be investigated whether the metabolic events which lead to positioning and retention of cells in one of the main periods of the mitotic cycle (G_1,S,G_2) in Van't Hof's pea root system[26,27] may apply to the meristems of ripening seeds.

REFERENCES

1. Alfert,M., and Geschwind, I.I.(1953) A selective staining method for the basic proteins of cell nuclei. Proc. Natl. Acad. Sci. 39:991-999.

2. Avanzi, S., Brunori, A., and D'Amato, F. (1969) Sequential development of meristems in the embryo of Triticum durum. A DNA autoradiographic and cytophotometric analysis. Developmental Biol. 20:368-377.

3. Avanzi, S., Brunori, A., D'Amato, F., Nuti Ronchi, V., and Scarascia Mugnozza, G. T. (1963) Occurrence of 2C (G_1) and 4C (G_2) nuclei in the radicle meristems of dry seeds in Triticum durum. Its implications in studies on chromosome breakage and developmental processes. Caryologia 16:553-558.

4. Avanzi, S., Brunori, A., and Giorgi, B. (1966) Radiation response of dry seeds in two varieties of Triticum durum. Mutation Res. 3:426-437.

5. Bogdanov, Y. F., and Jordansky, A. B. (1964) Autoradiographic investigations of nuclei of root meristem of germinating pea seeds with the employment of H^3-thymidine. Zh. Obshch. Biol. 25:357-363.

6. Bogdanov, Y. F., Liapunova, N. A., and Sherudilo, A.I. (1967) Cell population in pea embryos and root tip meristem. Microphotometric and autoradiographic studies. Tsitologia 9:569-576.

7. Brunori, A. (1967) Relationship between DNA synthesis and water content during ripening of Vicia faba seed. Caryologia 20:333-338.

8. Brunori, A., and Ancora, G. (1968) The DNA content of nuclei in the embryonic root apices of Allium cepa and their radiation response. Caryologia 21:261-269.

9. Brunori, A., Avanzi, S., and D'Amato, F. (1966) Chromatid and chromosome aberrations in irradiated dry seeds of Vicia faba. Mutation Res. 3:305-313.

10. Brunori, A., and D'Amato, F. (1967) The DNA content of nuclei in the embryo of dry seeds of Pinus pinea and Lactuca sativa. Caryologia 20:153-161.

11. Brunori, A., Georgieva, J., and D'Amato, F. (1970) Further observation on the X-irradiation response of G_1 cells in dry seeds. Mutation Res. 9:481-487.

12. Bryant, T. R. (1969) DNA synthesis and cell division in germinating onion. I. Onset of DNA synthesis and mitosis. Caryologia 22:127-138.

13. Bryant, T. R. (1969) DNA synthesis and cell division in germinating onion. II. Mitotic cycle and DNA content. Caryologia 22:139-148.

14. Caldecott, R. S., and Smith, L. (1952) A study of X-ray induced chromosomal aberrations in barley. Cytologia 17:224-242.

15. D'Amato, F. (1964) Nuclear changes and their relationships to histological differentiation. Caryologia 17:317-325.

16. D'Amato, F. (1965) Endopolyploidy as a factor in plant tissue development, in 'Proc. Internatl. Conference on Plant Tissue Culture' (ed. P. R. White and A. R. Grove), McCutchan Publ. Corp., Berkeley, Calif. :449-462.

17. Deitch, A. D. (1955) Microspectrophotometric study of the binding of the anionic dye, naphtol yellow S, by tissue sections and purified proteins. Lab. Invest. 4:324-351.

18. Dubinin, N. P., Nemtseva, L. S., and Halikov, P. H. (1971) Radiation induced chromatid aberrations in the G_1 phase of Crepis capillaris cells. Caryologia 24:117-128.

19. Evans, H. J. (1965) Effects of radiations on meristematic cells. Radiation Botany 5:171-182.

20. Evans, H. J., and Savage, J. R. K. (1963) The relation between DNA synthesis and chromosome structure as resolved by X-ray damage. J. Cell Biol. 18:525-540.

21. Grant, C. J. (1965) Chromosome aberrations and the mitotic cycle in *Triticum* root tips after X-irradiation. Mutation Res. 2:247-262.

22. Hsu, T. C., Dewey, W. C., and Humphrey, R. M. (1962) Radiosensitivity of cells of Chinese hamster *in vitro* in relation to the cell cycle. Exptl. Cell Res. 27:441-452.

23. Röbbelen, G., and Nirula, S. (1965) Der Replikationzustand der chromosomalen DNS in ruhenden Samen von *Haplopappus gracilis*. Naturwiss. 52:649-650.

24. Sire, M.W., and Nilan, R. A. (1959) The relation of oxygen post-treatment and heterochromatin to X-ray-induced chromosome aberration frequencies in *Crepis capillaris*. Genetics 44:124-136.

25. Stein, O.L., and Quastler, H. (1963) The use of tritiated thymidine in the study of tissue activation during germination in *Zea mays*. Amer. J. Bot. 50:1000-1011.

26. Van't Hof, J. (1968) Control of cell progression through the mitotic cycle by carbohydrate provision. I. Regulation of cell division in excised plant tissue. J. Cell Biol., 37:773-778.

27. Van't Hof, J. (1970) Metabolism and the prolonged retention of cells in the G_1 and S period of the mitotic cycle of cultured pea roots. Exptl. Cell Res. 61:173-182.

28. Wangenheim, K. H., and Walther, F. (1968) Strahlenbiologische Untersuchungen an Getreidearten. II. Unterschiedliche Strahlensensibilität bei gleichem Kernvolumen und DNA-gehalt. Radiation Botany 8:251-258.

29. Wolff, S., and Luippold, H. E. (1964) Chromosome splitting as revealed by combined X-ray and labelling experiments. Exptl. Cell Res. 34:548-556.

-DISCUSSION-

CONGER, B. - The ratio of 2C and 4C cells is quite different in different meristems of the same plant. Would you anticipate a corresponding yield of chromosome versus chromatid aberrations?

D'AMATO - Yes. In *Triticum* the correspondence is exact.

CONGER, A. - Have you measured total protein content in different meristems?

D'AMATO - No. The analyses have only been done for root meristems.

VAN'T HOF - There are two analogous situations relevant to this discussion. Firstly, the environmental conditions of the preceding growing season can affect the accumulation of a few cells in S and in the laboratory similar results can be obtained with stationary phase meristems. If such meristems are incubated with sucrose for a short period of time, say eight hours, about six percent of the cells will enter S. Depriving the cells of oxygen at this time causes them to remain in S until oxygen is again supplied. The cells will then advance to G_2 and stop a second time if no sucrose is provided after termination of hypoxia; if sucrose is provided, the cells will progress through G_2, M, etc., unimpeded. One can, therefore, "spurt" cells through the cycle. The ability to stop cells in S with hypoxia illustrates that the replication of DNA, as with other tested organisms, is an energy requiring process.

Secondly, Marcus and Feeley have shown with peanut cotyledons and wheat embryos obtained from seed that the percent of polysomes increases with imbibition indicating that, while dormant, seeds have fewer polysomes (J. Biol. Chem., 240:1675-1680, 1965). Similarly, carbohydrate starvation greatly reduces the percent polysomes in excised root meristems and the rate of ribosome dissociation corresponds well to the rate at which the mitotic activity ceases. Further, the percentage of polysomes will remain very low until carbohydrate is provided. The ribosomes then reassociate and mitotic cycle activity resumes. There is not, however, a close correspondence between the increased polysome percentage and mitotic index after carbohydrate replenishment.

CONGER, A. - Only a small fraction of the total protein in the nucleus is on the chromosome. Do you have any idea, Dr. D'Amato, how total nuclear protein relates to chromosome aberrations?

D'AMATO - We have no idea. In *Helianthus* there is a considerable variation in nuclear volume among 2C cells. The nuclear volume does not necessarily reflect the DNA value -- some 2C nuclei are very large. It might be that total protein relates significantly to nuclear volume.

MORPHOGENESIS OF PRIMORDIA OF LATERAL ROOTS

D. DAVIDSON

Department of Biology, McMaster University

Hamilton, Ontario, Canada

INTRODUCTION

When an angiosperm seed germinates the radicle of the embryo grows out and develops into the primary root; this later becomes a complex structure by giving rise to lateral, or secondary, roots. The emergence of the radicle and the subsequent growth of the primary root show a marked polarity: new cells generated by mitosis and the new growth produced by cell elongation are added almost exclusively along one axis. The primary root shows longitudinal growth. It may be inferred from this observation that 1) the pattern for the future growth of the primary root is established during the period of embryogenesis and 2) the internal physiological conditions necessary to realize this imprinted growth pattern exist in the embryo or are established under normal conditions of germination. All primary growth of the primary root that occurs after radicle emergence perpetuates the initial polarized growth along one axis and the root has the form of a long cylinder with a conical apex.

Lateral roots are not present during the period of embryo growth; they arise _de novo_, sometime after radicle emergence, from cells lying deep in the tissue of the primary root. While it is still embedded in the tissues of the primary root the group of cells that will eventually emerge as a lateral root is called a primordium. In some species primordia form only from the pericycle, in other species both the pericycle and endodermis contribute to a primordium. Polarity is established in the primordia within 2-3 days and eventually it grows mainly along one axis, in a manner similar to the primary root. We shall be concerned here with: 1) the initiation, growth, development and emergence of the primordia. 2) the organization of the apical meristem of the lateral root. 3) the physiological changes that occur in the meristematic cells as a primordium becomes a lateral root. 4) the evidence bearing on hormonal control of cell proliferation in the apical meristem. The experimental results presented here are mainly from studies of *Vicia faba*. The techniques for growing roots and staining and preparing squashes have been described (Davidson[8]).

RESULTS and DISCUSSION

INITIATION OF PRIMORDIA

A primordium is initiated when a group of pericycle and/or endodermal cells around a xylem pole begin to proliferate. As cell number increases and they grow, they form a rounded bump that grows away from the central cylinder and towards the epidermis. Its axis of growth lies at right angles to the long axis of the primary root. Divisions in the very young primordia are oriented periclinally and anticlinally and the primordium increases in diameter as well as in length. The planes of cell division are much more random than in a primary root meristem. Thus, there is no evidence, in the young primordium, of the type of organized growth seen in a primary root that is indicative of an internal polarity; that is established later.

A lateral root primordium of *V. faba* grows from a number of initial cells, not from a single cell. Stable, visible chromosomal rearrangements e.g. reciprocal translocations (Davidson[4,5,6,7,8]) and colchicine induced polyploidy (Davidson[8,9]) act as cell markers and they have been used to estimate the number of cells from which a primordium develops. These estimates are made in the following way. Following irradiation or colchicine treatment newly initiated primordia and laterals include cells with changed karyotypes. The number of different karyotypes present in the meristem is taken to be the absolute minimum number of cells that acted as primordium initials. The relative frequencies of the different karyotypes provides an estimate of the actual minimum number of initial cells.

For example, a primordia with 4 different karyotypes must have arisen from a minimum of 4 cells. But if the relative frequencies of these 4 cell types are 2:4:7:1 then it is concluded that 14, i.e. 2+4+7+1, initial cells were involved. From a number of determinations the number of initial cells was found to range from 6 to 50 and the average number was calculated to be 24 (Davidson[8,9]). These results imply that 1) cells descended from several different cell lineages occur together in the small region of the pericycle that would be made up of 24 cells, 2) in meristems that are cytological chimaeras the different cell types are mixed together i.e. all the cells with the same karyotype are not confined to a particular segment of the meristem. The mixoploid composition of the meristems of lateral roots that develop on a regenerated primary root or on a lateral root growing from the C-tumor reveal that the pericycle from which they were derived was mixoploid (Fig. 1).

GROWTH OF PRIMORDIA

The time from the initiation of a primordium to its emergence as a lateral root is about 5 days. This estimate was obtained by inducing marker cells in incipient primordia and timing the emergence of marked primordia. Treatments with weak solutions of colchicine, 0.003% for 3 hours, have almost no effect on primary root growth i.e. the C-tumor formed shows a minimal amount of swelling and elongation is only slightly inhibited. Pericycle cells, however, are sensitive to this concentration of colchicine and tetraploid and octoploid cells are formed. Friedberg and Davidson[20] used the presence of 4n and 8n cells as a marker to time the duration of growth of a primordium; in the course of their study they found that a primordium consists of two sub-populations of cells that are spatially discrete and that show changes in the level of mitotic activity

Primary roots were treated with 0.003% or 0.006% solutions of colchicine and the frequencies of 2n, 4n and 8n mitoses were determined daily from 2 to 8 days after treatment. Upto the time when a very large primordium has formed i.e. about 4 days, 4n cells made up 60% or more of the mitoses. Thus, whether their frequency is expressed as a percentage of mitoses or as mean frequency per 1000 cells, the 4n cells make up the bulk of cells seen in mitosis. Within a further 24 hours,

Fig. 1. Lateral roots and the composition of meristems in *V. faba*. These roots developed after treatment with 0.05% colchicine for 3 hr. The root system was fixed 9 days after treatment. Data from Davidson[9], Tables 3 and 4; all data from one root system.

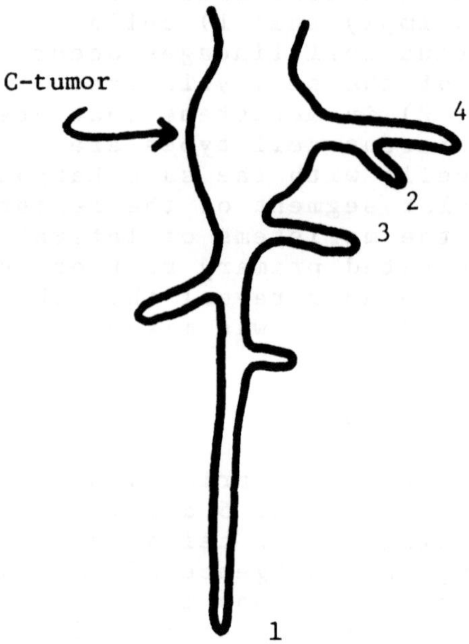

2. a lateral root from C-tumor

4. lateral root growing from a C-tumor lateral root

3. lateral root growing from the regenerated portion of the primary root

1. regenerated primary root

	Ploidy of cells at metaphase				
	2n	4n	8n	No. of cells	% polyploid cells
primary root	122	38	-	160	23.7
C-tumor lateral	35	91	34	160	78.1
lateral on 1° root	35	81	4	120	70.8
lateral on C-tumor lateral	14	84	2	100	86.

however, the frequency of 4n cells drops and the frequency of 2n mitoses increases. Once the primordium has emerged the meristem reverts to its original composition i.e. 4n cells increase and 2n cells decrease in frequency. An examination of sections of primordia showed that the 4n cells formed a central core and the 2n cells lay around the core, as a sheath. Furthermore, a microspectrophotometric study revealed that when the 4n cells undergo this temporary decrease in mitotic activity they are held in G_1. Thus, for a period of 24-36 hours, just before emergence, the central core of cells do not progress through interphase but are blocked in G_1. After emergence, this block is reversed, the central cells proliferate again and give rise to the meristem of the young lateral root (Friedberg and Davidson[20]).

This change in the organization of the primordium is interesting for the following reasons: 1) a block in DNA synthesis is temporarily imposed on a spatially discrete group of cells while adjacent cells undergo S and proliferate actively. 2) this block occurs at a specific stage in development i.e. it occurs at about the time of emergence and when the peripheral cells of the primordia are still proliferating and are forming a root cap. 3) the change in the locus of active cell proliferation is superimposed on a population of cells that is heterogeneous for cell cycle times and which, in the course of about 6 days, undergoes a number of striking physiological changes. These are summarized in Table 1. In particular, the change in the ability of primordia to incorporate ^{3}H-TdR has not been explained satisfactorily. Large primordia do take up ^{3}H-TdR since Webster (unpublished) has found cytoplasmic labeling after exposure to ^{3}H-TdR but for some reason it is not used in DNA synthesis. Results obtained from a study of the response of large and very large primordia to FUdR suggest that the failure to incorporate ^{3}H-TdR is not due to the absence of thymidine kinase. Like the differential response of primordia and lateral roots to X-rays (Howard[21]) the significance of these physiological changes for the morphogenesis of the primordium meristem is unknown. Nor do we have any indication of the possible relationship of one change to another. In the primordium of *V. faba* we have a population of proliferating cells that, while increasing in number, simultaneously undergo a series of physiological changes.

Other differences are found between cells of primordia and of fully emerged laterals. The length of

Table 1. Normal changes and response to various agents of cells of small primordia (SP), large primordia (LP), very large primordia (VLP), just not emerged primordia (JNE), and lateral roots of V. faba. References in parentheses.

1. Incorporation of ^{3}H-TdR into DNA
   - SP: low initially; labeling index rises to ∿50% after 12 hrs.
   - LP, VLP and JNE: little or no incorporation LI is < 1.
   - Lateral: rapid incorporation (15, 22, 26, 27)

2. Incorporation of ^{3}H-AdR and ^{3}H-CdR into DNA
   - SP: low initially; increases with time cf. ^{3}H-TdR.
   - L: immediate incorporation: LI varies with precursor. (27)

3. Response to colchicine (0.025%)
   - SP: sensitive; 4n cells form; growth suppressed.
   - LP, VLP: less sensitive than SP or L; 4n cells formed but growth is not inhibited by 0.025% solutions.
   - JNE: 4n cells formed; growth in inhibited.
   - L: sensitive; 4n cells formed; growth is inhibited. (10, 11, 12)

4. Percentage slow cycling cells determined by 5-aminouracil treatment
   - LP: 18.1; VLP: 16.4; L: 17.8 (30)

5. Response to IAA
   - LP, VLP and JNE: emergence and growth is not prevented by IAA.
   - L: IAA inhibits growth; many roots are killed; S is delayed. (14, 18, 35)

6. Response to 500 ppm 5-aminouracil
   - SP: sensitive; mitotic index reduced; recovery begins after 9 hours.
   - LP, VLP, JNE: less sensitive than SP or L; mitotic index not reduced to the same extent as in L; recovery begins after 9 hours.
   - L: sensitive; mitotic index much reduced; recovery begins after 12 hours. (29, 32)

7. Mean cycle time of fast cycling cells
   - SP: 12 hours
   - LP: ∿12 hours
   - L: 14 hours (26, 34)

mitotic cells is about the same in large primordia and lateral roots (Fig. 2) though the relative frequencies of cells of different sizes was somewhat different; large primordia had relatively fewer large cells. In diameter, however, cells of large primordia ranged from 11-25 μ and had a mean value of 18.2±3.4 μ compared with a range of 7-20 μ and a mean diameter of 12.2±3 μ in lateral roots.

Nuclear diameter has also been determined for cells of large primordia and lateral roots (Fig. 3). Nuclei of diameter 7 and 8 μ are present in laterals but not in primordia. Though the mean nuclear diameters are almost identical, 11.4 μ in laterals and 11.9 μ in large primordia, the frequency distributions of different size classes on nuclei differ in the tissues (Fig. 3). Apart from the relatively higher frequency of nuclei with a diameter of 12 or 13 μ in the large primordia they also have a low frequency, 1.2%, of cells with elongated nuclei; in laterals there are 8.99% cells with such nuclei. A plot of the cumulative frequency of each nuclear volume class against the log of nuclear volume reveals that in large primordia all the cells with spherical nuclei are part of a single population. In lateral roots it appears that, on the basis of nuclear volume, there are two populations. Thus, in the transition from a large primordium into a lateral root in addition to the physiological changes, the morphology of the nuclei also changes: the range of nuclear diameters increases and the frequency of cells with elongated nuclei also increases. These changes in nuclear and cell size raise several questions that apply to all root meristems and that are particularly crucial for meristems in which there is marked heterogeneity for cell cycle duration. They are: 1) are nuclear size and the duration of S or of the cell cycle directly related? 2) do the rates of increase in nuclear size differ in cells having different cycle times? 3) is the rate and amount of cell growth related to the duration of the cell cycle? In considering these questions we turn first to the distribution of mitoses in a meristem.

DISTRIBUTION OF MITOSES

Roots of V. faba were hydrolysed in 1 N HCl for 10 minutes, stained with Feulgen and columns of cells were separated from one another under a dissecting microscope using fine needles. To help in identifying cell boundaries light green was added as a counterstain.

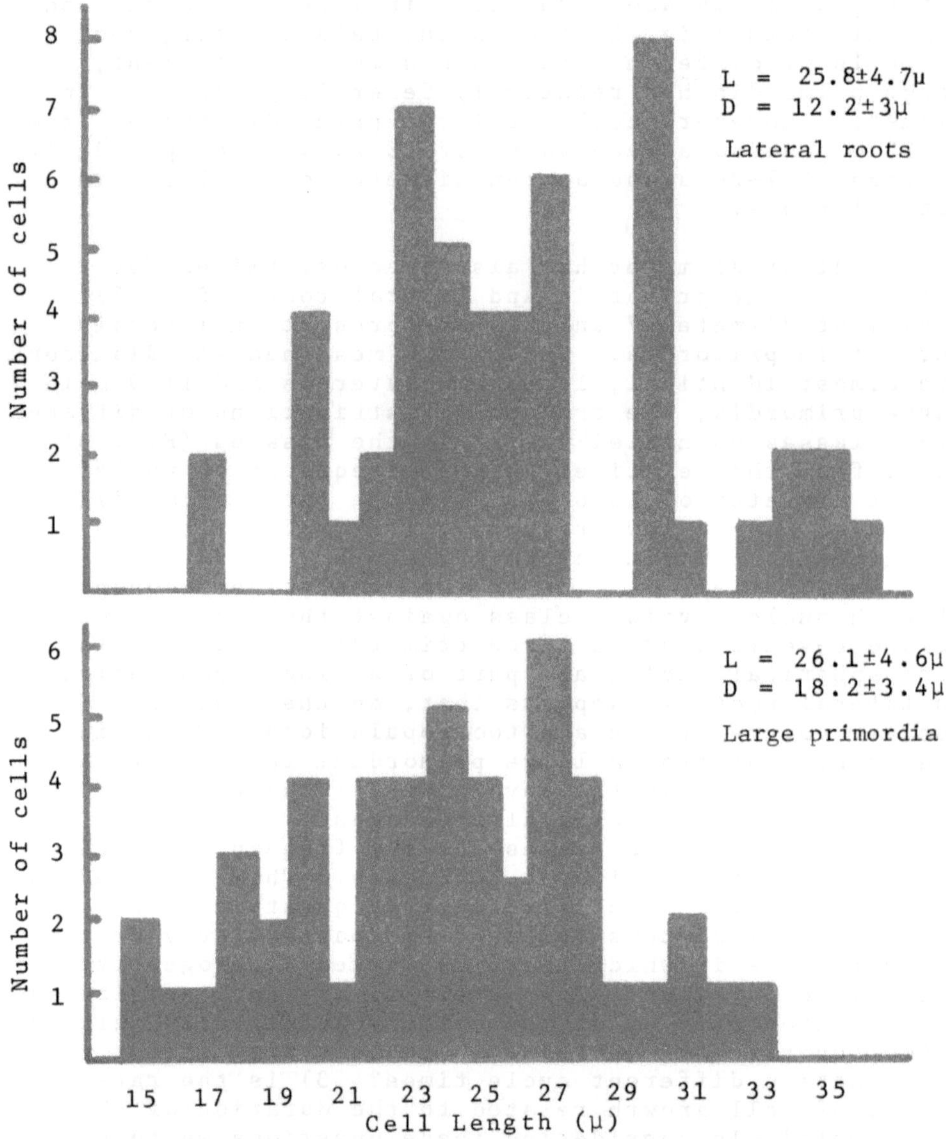

Fig. 2. Mean cell length ± S.D. (L) and mean cell diameter ± S.D. (D) of 50 cells in large primordia or lateral roots. The samples were made up as follows: large primordia 23 prophases, 13 metaphases, 11 anaphases and 3 telophases; laterals 27P, 9M, 11A and 3T. Measurements were made on meristems that had been hydrolysed, stained with Feulgen's reagent and counterstained with light green. They were not squashed.

The cells were not squashed. The frequency with which 1,2,3 or 4 mitoses occurred in cell adjacent to each other along a column of cells was determined (Table 2).

Table 2. Frequency of mitoses occurring singly or adjacent to another cell in mitosis. The distribution of mitoses was scored in columns of cells dissected from lateral root meristems of V. faba.

1 mitosis: cell on either side was in interphase	mitoses in adjacent cells 2	3	4	Total mitoses	Total cells
117	27	3	2	188	506

Most mitoses occurred singly, with the two cells on either side in interphase; this was the situation for 62.2% of mitoses scored. Only 54 mitoses were in pairs, i.e. 27 pairs of longitudinally adjacent cells in mitosis, and of these only 3 pairs showed synchrony of mitotic stage. And since mitosis lasts about 2 hours, even in pairs of cells both of which are in mitosis, the degree of synchrony is really less than appears from the values in Table 2. The total number of cells scored in this sample was 506 i.e. mitoses and the interphase cells on either side of all mitotic cells. Mitotic index for this sample is 37.15. It is higher than the MI of 8.1 for the whole meristem because cells in mitosis were selected for. Now if there was a tendency for a cell in mitosis to be next to another mitotic cell, then the probability of finding two adjacent cells in mitosis would be 0.14. The observed frequency is 0.12. Thus, the frequency expected by chance is actually slightly higher than the observed frequency. It appears that there is little, if any, tendency for cells in mitosis to occur in pairs along a column of cells. This observation has implications for the fast cycling cells of a lateral root meristem. We have seen that 50% of the cells have a cell cycle duration of 14 hours and these cells make up about 75% of the cells in mitotis at any time. Therefore, of the cells seen in mitosis the majority must be longitudinally adjacent to their sister cell. Since the frequency of pairs of cells in mitosis is so low we must conclude that: 1) two sister-cells that enter G_1 together rapidly lose their synchrony and they progress through the cell cycle at different rates; 2) if a growth hormone that controls the rate of progress through the

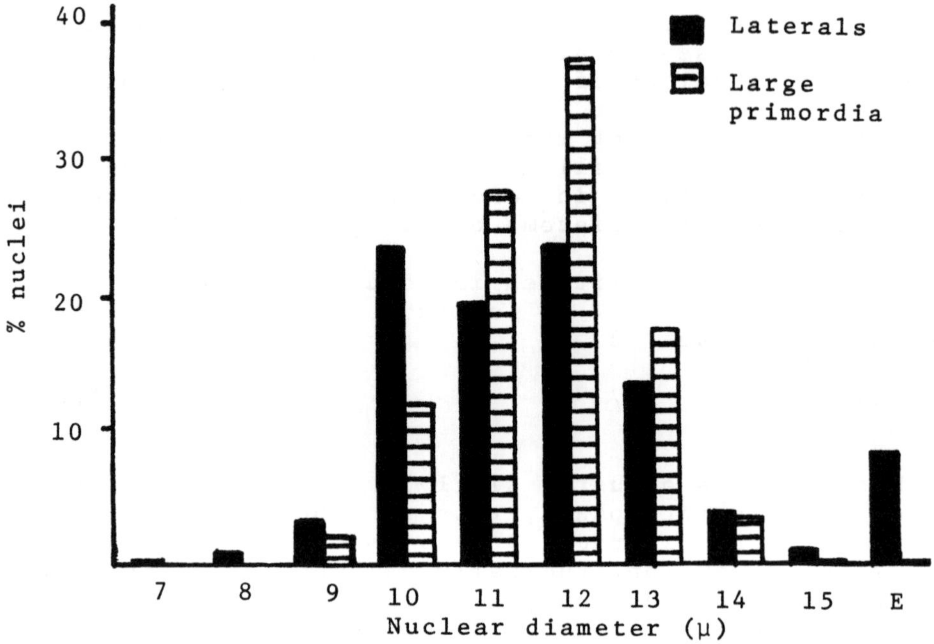

Fig. 3. Nuclear diameter in cells of lateral roots and large primordia. 489 nuclei were measured in laterals and 487 in large primordia. Most nuclei closely approximate spheres and only 1 diameter was measured but cells with elongated nuclei were also present and have been included in the histogram (E).

cell cycle moves in a polar fashion along a column of cells, then it does not maintain even a reasonable degree of synchrony of growth between longitudinally adjacent sister-cells. Further evidence on these points comes from a study of the pair of sister-cells produced by an anticlinal division and from cells that are laterally adjacent i.e. they are equidistant from the root apex and lie in adjacent columns of cells.

Columns of cells were dissected out from lateral roots as before. Cells were scored as being adjacent if their walls overlapped by at least 4 μ. Of 109 pairs of laterally adjacent cells scored, 100 pairs had only 1 cell in mitosis and 9 pairs had both cells in mitosis. Of these 9 pairs, 6 were non-sisters and 3 were pairs of sister-cells. Only 1 pair of dividing cells showed stage synchrony; they were at metaphase. In pairs of adjacent cells, mitosis in one cell is accompanied by mitosis in the other cells in only 8.3% of the cases.

There is no obvious tendency for pairs of adjacent cells to divide more or less in synchrony.

Additional pairs of cells formed by an anticlinal division were scored. In 26 pairs, both cells were in interphase; in 16 pairs one cell had divided and the other had not; and in 3 pairs both cells were in mitosis. The conclusion from these results is that in both linear and lateral pairs of cells synchrony is rapidly lost between sister-cells.

Loss of synchrony can apply to two cells that are members of the same population or to two cells that diverge in rate of progress through the cell cycle to such an extent that one cell becomes slow cycling or non-dividing while the other remains fast cycling. The loss of synchrony can be due to an internal factor or some control exerted on cells by some integrating growth factor moving within the meristem. And it may occur in localized regions of the meristems or it may occur independently at different points along a column of cells. Since we have no evidence on the functions of growth factors in this respect we must look for other possible causes.

The two sister-cells produced by a mitosis are genetically equivalent. Cytokinesis does not produce two cells of equal volume in all cases, however. Only 82.9% of divisions produce two cells that at least appear to be equivalent; the others produce two cells that appear to differ in size (Fig. 4). The production of sister-cells of unequal sizes may contribute to the change from a fast cycling to a slow cycling cell. And in the case of triangular cells it may be involved in the pattern of cell elongation in different parts of a column of cells. An unequal distribution of cytoplasmic components between sister-cells may account in part for variations in cell cycle duration but it is probably not the only determining factor. We can now consider the distribution within the meristem of cells having different cycle times.

The relatively constant ratio of fast cycling to slow cycling cells in a meristem can be maintained only if a proportion of fast cycling cells become slow cycling. Clearly a group of fast cycling cells that divide every 14 hours would rapidly overgrow the slow cycling cells. Thus, the two sister-cells produced by a mitosis of a fast cycling cell do not remain fast cycling; some must change into slow cycling or even

200 7 6 24 4

Number of anaphases and telophases

Fig. 4. Anaphase and telophase configurations and their frequency in normal lateral root meristems of V. faba.

non-dividing cells.

In analyzing the positions of fast and slow cycling cells in the meristem two approaches were used. One used the technique developed by Socher and Davidson[31] using 5-aminouracil. Cells in G_2 continue to proceed to mitosis in the presence of 5-AU; cells at the S-G_2 transition, or perhaps at the very beginning of G_2, are held there. The mitotic index drops when roots are treated with 5-AU. The first cells to be affected are fast cycling cells and by 6 hours the only cells entering mitosis in laterals of V. faba are slow cycling cells (Socher and Davidson[30,31,32]). Longitudinal sections were made of roots treated with 5-AU for 12 hours. The mitotic index was low since only slow cycling cells were dividing; but these mitoses were distributed throughout the meristem. We conclude that slow cycling cells are not confined to one part of the meristem and, therefore, the transition from the fast to the slow cycling condition occurs throughout the columns of cells making up the root (Davidson and Socher; unpublished).

The second approach was to treat roots with 0.025% colchicine for 3 hours and determine the distribution along columns of cells of diploid and tetraploid mitosis 15 to 21 hours from the beginning of treatment. Tetraploid cells in division at 15 hours have completed 2 mitoses in that time and can be considered to be fast cycling cells; diploid cells, on the other hand, are undergoing their first mitosis since the treatment and they must be mainly slow cycling cells.

Roots were hydrolysed and stained and columns of cells were dissected out. This is easiest in 15 hour fixations; by 21 hours the lateral expansion of the cells appears to disrupt the binding of cells and they tend to separate into small groups or into individual cells. Diploid and tetraploid cells, in division, were distributed along the columns of cells (Fig. 5). The results of this study agreed with those from the 5-AU treatment and they agree with the results of Rasch, Rasch and Woodard[28] using ^{3}H-TdR: fast and slow cycling cells are distributed, apparently in a random manner, along individual columns of cells. Furthermore, of 2 laterally adjacent cells i.e. from two adjoining columns, one may be fast cycling and the other slow cycling (Fig. 5C). An hypothesis that invokes a growth factor that is 1) exported into the meristem from one control center and 2) moves in a polar fashion appears to be incapable of explaining the distribution of cells having different cycle times either along one column of cells or at identical points in two adjoining columns. The control of cell cycle duration appears to be either an inherent property of the cells or to operate only over small distances within each column. Furthermore, if cell elongation is controlled by a hormone that is distributed in a strictly polar fashion we would expect laterally adjacent cells to increase in length at identical rates; they may not always do so (see next section).

RESPONSE TO COLCHICINE

When diploid and tetraploid cells are found together in division 15 hours after the start of a 3 hour treatment with colchicine it may appear that there is, in fact, little difference in their cycle times. However, it must be remembered that one response to colchicine is a speeding up of the rate of progress through the cell cycle (Webster and Davidson[35]). G_1 cells are induced to reach mitosis much earlier than they would do so in an untreated lateral. This response is, in large part, the cause of the dramatic rise in mitotic index seen several hours after treatment with colchicine (MacLeod and Davidson[25]; Webster and Davidson[35]). This stimulation to reach mitosis faster affects mainly slow cycling cells. The result is that while in a 14 hour period in a normal lateral root there is 1 mitosis for approximately every 2 cells, in a colchicine treated root there is 1 mitosis for every cell in the meristem. This does not mean that every cell has divided; the 4n cells seen in mitosis are undergoing their second division in 14 hours. But

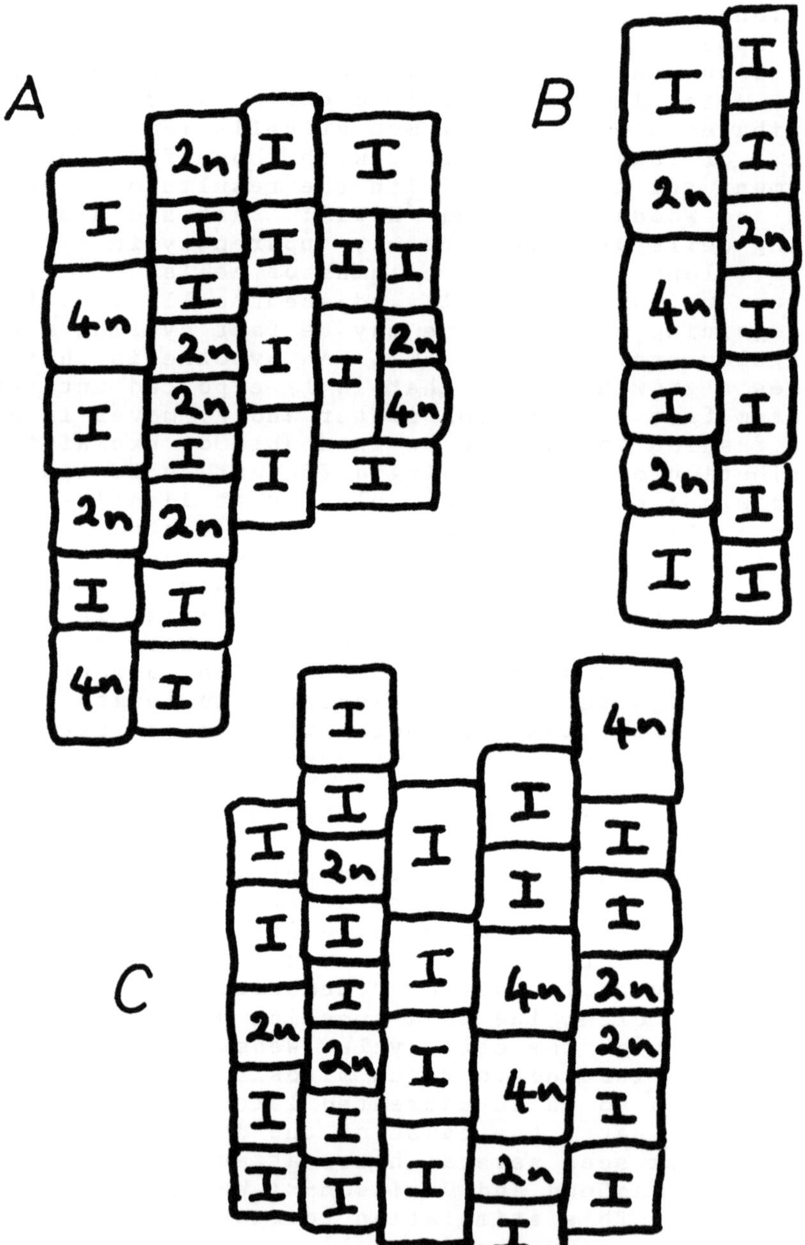

Fig. 5. Distributions of tetraploid (4n) and diploid (2n) mitosis and interphase (I) cells in columns of cells dissected out of lateral roots of V. faba treated with 0.025% colchicine for 3 hours and fixed 15 hours from the beginning of treatment.

even before the 4n cells appear in mitosis the mitotic index increases (MacLeod and Davidson[25]; Webster and Davidson[35]); this rise is due solely to a reduction in the duration of the cell cycle of the diploid cells. Thus, the overall increase in mitotic activity shows how effectively colchicine stimulates progress through the cell cycle. And the fact that 2n cells continue to be seen in mitosis 20-24 hours after colchicine treatment is further evidence of the presence of slow cycling cells.

Colchicine also affects cell expansion. The so-called C-tumor that forms in the region basal to the meristem is produced by the isodiametric expansion of the cells i.e. they expand in diameter as well as length, rather than elongating along one axis. A similar effect occurs in the meristem in the tetraploid cells. In lateral roots for example, 15 hours after the start of a 3 hour treatment with 0.025% colchicine the tetraploid cells in division were 45.1±6.6 μ in length and 21.6±4.2 μ in diameter (30 cells scored); diploid cells were 24.2±7 μx 13.9±3.9 μ (100 cells scored). In large primordia the 4n cells were less affected by colchicine. They were 31.2±8.1 μ x 18.9±4.5 μ (50 cells) while the diploid cells were 23.4±4.4 μ x 15.3±3.2 μ (50 cells). The increase in diameter of the 4n cells must induce localized disruption of cell-cell contacts and presumably it produces some disorganization of the internal architecture of the root. Considered solely from the point of view of cell expansion and its control, the increase in cell size of 4n cells in the interphase that follows 4n cell formation poses a problem: how can a 4n cell double its length (Fig. 5C) while laterally adjacent cells do not also double in length? In posing the question it is presupposed that: 1) there is no sliding growth within a meristem; 2) an increase in cell length in an cell, X μ from the apex is accompanied by an exactly equal increase in cell length in all other cells that lie X μ from the apex. In fact, however, some sliding growth may occur, though it would be of limited extent: it may be correlated with the presence of the triangular or dome shaped cells that occur at intervals along a column of cells. The larger question raised here concerns the rate at which increase in cell length occurs in different parts of a meristem. There is no experimental evidence that enables us to say whether all cells in a meristem increase in length at the same rate, whether the rate of increase differs in different parts of meristem or whether growth is a periodic phenomenon.

Finally it should be noted that 4n cells in large primordia are smaller than those in laterals. Their dimensions are 31.2±8.1 μ x 18.9±4.5 μ (50 cells); their response to colchicine is less marked than that of cells of lateral root meristems (Davidson[10,11,12]; Table 1).

REGULATION OF CELL PROLIFERATION AND GROWTH

Meristems are highly organized structures. The polarity of elongation and cell division, the relatively constant ratio of fast and slow cycling cells and the maintenance of a cell cycle of ∿14 hours in both 2n and 4n cells in a mixoploid lateral root (Friedberg and Davidson[19]) all indicate a strict regulation of all these facets of growth. The regulatory mechanisms must act as inhibitors as well as stimulators since neither cell proliferation nor cell expansion occur at random sites nor at random rates within the meristem. To preserve the form of a growing root at least one mechanism that integrates all growth into a unified whole must be present. One control mechanism may regulate all aspects of growth; or it may integrate the various types of growth, each of which has localized regulatory systems distributed throughout the meristems. As we have seen, the random distribution of fast and slow cycling cells along columns of cells strongly suggest the existence of control systems that extend only over small distances in cell columns. Furthermore the presence of a quiescent center, in which mitoses are infrequent and cell cycle duration is longer than in many other cells of the meristem (Clowes[3]) is also indicative of a local control of the rate of cell proliferation.

The obvious candidates to fulfill the postulated role of controllers of growth are the well known growth factors, auxins e.g. IAA, gibberellins and cytokinins. In a number of plant tissues, particularly cambium, IAA induces cell division. But it does not do so in all tissues, suggesting that auxin "..is but one of several potentially limiting growth factors' (Thiman[33]). It is relevant here to point out that the stimulation of cells to divide has not yet been shown to be a specific property of IAA i.e. that the cells have a specific requirement that can be satisfied only by IAA. Nevertheless exogenous IAA can stimulate growth and maintain mitotic activity in meristems previously treated with colchicine at the time when mitotic index would normally fall (Davidson and MacLeod[13];

Davidson, MacLeod and Taylor[16]; MacLeod and Davidson[24]). This evidence suggests that IAA can contribute to the maintenance of the proliferative state; it does not, however, show that IAA is specifically required or that the auxin controls the rate of proliferation or the axes of division.

Evidence for the polar transport of a growth factor exists only for IAA and that evidence has been obtained from coleoptiles and shoots. In roots, as Audus[1] points out, applied IAA is not transported in a polar manner from the apex of the root to the zone of cell extension; instead it tends to accumulate in the meristem. In this respect the behaviour of roots does not agree with that of coleoptiles and shoots. The auxins present in roots may differ from those in shoots. In _V. faba_ roots it appears that the main auxin is not IAA (Burnett, Audus and Zinsmeister[2]). These studies suggest that perhaps our view of the extent of the role auxins play in the control of root growth has been prejudiced by the results from studies of coleoptile and shoot growth. If cell proliferation and cell elongation are controlled by auxin then the two processes are affected to different extents. In irradiated roots of _V. faba_, for example, mitosis continues even though the roots are not elongating (Davidson[4,8]) suggesting that the factors responsible for elongation are not completely responsible for cell division. In onion roots showing a geotropic curvature it had been claimed that the cause of the curvature was differential rates of cell division in the upper and lower halves of the root. This claim is statistically unfounded (Audus[1]), which means that differential elongation, presumably due to a differential supply of some auxin to the two halves of the root, occurs without a corresponding differential rate of cell proliferation. And in _Zea_ roots, removal of the root cap resulted in a loss of sensitivity to a geotropic stimulus though not in a loss of ability of the cells to elongate (Juniper et al.[23]). This is further evidence that the growth that occurs in a tropic response is not controlled by the system responsible for cell elongation during straight growth.

CONCLUSION

The meristem of a lateral root has a complex organization. It is formed from a primordium that is also complex, though not identical physiologically with the lateral meristem. And the primordium, in turn, is produced by a group of cells in the pericycle that differ in some respects from the cells of the primary meristem from which they were derived (Davidson, MacLeod and Webster[17]). The increase in size of these proliferating populations of cells is accompanied by a surprising number of physiological changes. In particular, in the transition from a very large primordium to a newly emerged lateral root several changes occur that suggest the activation of different genes. Our understanding of the morphogenesis of mitotically active cells will remain incomplete until we have determined the role growth hormones play in this process. It is the evidence bearing on this aspect of growth that is particularly incomplete.

LITERATURE CITED

1. Audus, L. J. (1969) Geotropism. In: Physiology of Plant Growth and Development ed. M. B. Wilkins. McGraw-Hill. London pp. 203-242.
2. Burnett, D., Audus, L. J. and Zinsmeister, H.D. (1965) Growth substances in the roots of _Vicia faba_ II. Phytochem _4_:891-904.
3. Clowes, F. A. L. (1961) Apical Meristems. Blackwell Scientific Publ. Oxford.
4. Davidson, D. (1959) A method for estimating mitotic rates in _Vicia_ roots after X irradiation. Brit. J. Radiol. _32_:612-614.
5. Davidson, D. (1959) Changes in the chromosome complements of cells of _Vicia faba_ following irradiation. J. Exptl. Bot. _10_:391-398.
6. Davidson, D. (1960) Meristem initial cells in irradiated roots of _Vicia faba_. Ann. Bot. N.S. _24_:287-295.
7. Davidson, D. (1961) Protection and recovery from ionizing radiation mechanisms in seeds and roots. In: Radiation Protection and Recovery ed A. Hollaender. Pergamon Press. pp. 175-211.
8. Davidson, D. (1961) Mechanisms of reorganization and cell repopulation in meristems in roots of _Vicia faba_ following irradiation and colchicine. Chromosoma _12_:484-504.
9. Davidson, D. (1965) Cytological chimaeras in _Vicia_ roots. Bot. Gaz. _126_:149-154.
10. Davidson, D. (1965) A differential response of meristems of _V. faba_ roots to colchicine. Ann. Bot. N.S. _29_:253-264.
11. Davidson, D (1969) The response of primordial cells of _Vicia faba_ to colchicine. Ann. Bot. _33_:245-251.

12. Davidson, D. (1969) The differential response of meristems of V. faba to colchicine. Caryologia 22:213-221.

13. Davidson, D. and MacLeod, R. D. (1966) Changes in mitotic indices in roots of V. faba I. Antagonistic effects of colchicine and IAA. Chromosoma 18:421-437.

14. Davidson, D. and MacLeod, R. D. (1968) Response of meristems to indoleacetic acid and colchicine: differences between primordia and root meristems. Bot. Gaz. 129:166-171.

15. Davidson, D. and MacLeod, R. D. (1968) Heterogeneity in cell behaviour in primordia of Vicia faba. Chromosoma 25:470-474.

16. Davidson, D., MacLeod, R. D. and Taylor, J. (1965) An IAA-induced stimulation of growth of roots of Vicia faba previously treated with colchicine. New Phytol. 64:393-398.

17. Davidson, D., MacLeod, R. D. and Webster, P. L. (1968) Labeling of pericycle cells with H^3-thymidine. Naturwissenschaften 55:351-352.

18. Davidson, D. and Webster, P. L. (1968) Effects of IAA and myo-inositol on mitotic cycles. Proc. 6th Intern. Cong. Plant Growth Subst. ed. F. Wightman and G. Setterfield. Runge Press. pp. 543-557.

19. Friedberg, S. H. and Davidson, D. (1970) Duration of S phase and cell cycles in diploid and tetraploid cells of mixoploid meristems. Exptl. Cell Res. 61:216-218.

20. Friedberg, S. H. and Davidson, D. (1971) Cell population studies in developing root primordia. Ann. Bot. (in press).

21. Howard, A. (1966) Effect of X-rays on the number and growth rate of lateral roots of Vicia faba. Inter. J. Rad. Biol. 10:83-93.

22. Hummon, M. R. (1962) The effects of tritiated thymidine incorporation on secondary root production by Pisum sativum. Amer. J. Bot. 49:1038-1046.

23. Juniper, B. E., Groves, S., Landau-Schachar, B. and Audus, L. J. (1968) Root cap and the perception of gravity. Nature 209:93-94.

24. MacLeod, R. D. and Davidson, D. (1966) Changes in mitotic indices in roots of *Vicia faba* L. II. Long term effects of colchicine and IAA. New Phytol. 65:532-546.

25. MacLeod, R. D. and Davidson, D. (1968) Changes in mitotic indices in roots of *Vicia faba* III. Effects of colchicine on cell cycle times. Exptl. Cell Res. 52:541-554.

26. MacLeod, R. D. and Davidson, D. (1968) Delayed incorporation of H^3-thymidine by primordial cells. Chromosoma 24:1-9.

27. MacLeod, R. D. and Davidson, D. (1970) Incorporation of ^3H-deoxynucleotides: changes in labeling indices during root development. Canad. J. Bot. 48:1659-1663.

28. Rasch, R. W., Rasch, E. M. and Woodard, J. W. (1967) Heterogeneity of nuclear populations in root meristems. Carylogia 20:87-100.

29. Rudolph, J. and Davidson, D. (1971) (in preparation).

30. Socher, S. H. and Davidson, D. (1970) Heterogeneity in G_2 duration during lateral root development. Chromosoma 31:478-484.

31. Socher, S. H. and Davidson, D. (1971) 5-aminouracil treatment. A method for estimating G_2. J. Cell Biol. 48:248-252.

32. Socher, S. H. and Davidson, D. (1971) (in preparation)

33. Thiman, K. V. (1969) The Auxins. In: The Physiology of Plant Growth and Development ed. M. B. Wilkins. McGraw-Hill. London. pp. 1-45.

34. Webster, P. L. and Davidson, D. (1968) Evidence from thymidine-^{3}H labeled meristems of *Vicia faba* of two cell populations. J. Cell Biol. 39:332-338.

35. Webster, P. L. and Davidson, D. (1969) Changes in the duration of the mitotic cycle induced by colchicine and indol-3yl-acetic acid in *Vicia faba*

roots. J. Exptl. Bot. 20:671-685.

I acknowledge the support of the National Research Council of Canada and the help, in the preparation of the manuscript of M. McDermott and M. Haight.

-DISCUSSION-

HABER - With symplastic growth, two neighboring cells will, despite differences in size, have the same specific growth rate in any given direction, provided there is no cell slippage. Do you observe cell slippage?

DAVIDSON - I have no evidence of cell slippage. One would have to study plasmodesmatal connections between cells.

TORREY - In the shoot apex compensation for increment in cell length is often accomplished in adjoining cells by mitotic activity.

DAVIDSON - We have observed when treating cells with colchicine and producing tetraploid cells that there is a burst of mitotic activity by the diploid cells. Perhaps the tetraploid cells "pressure" the diploid cells into growth via mitosis.

CONSTANTIN - Does acute exposure to ionizing radiation change the ratio of fast cycling to slow cycling cells?

DAVIDSON - I suspect that it would, but I have no data at the moment.

TORREY - Is there any information about the elongation rates of fast cycling and slow cycling cells?

DAVIDSON - None at present.

TORREY - When does the quiescent center develop in the lateral roots?

DAVIDSON - Just after emergence.

CLOWES - Would you suspect that the core cells are derived from the endodermis and the peripheral cells from the pericycle?

DAVIDSON - This is what we suspect.

CLOWES - This would indicate then that the endodermal derived cells are programmed differently from the pericyclically derived cells.

CELL DIVISION, DEVELOPMENT, AND RADIATION INJURY*

ALAN H. HABER

Biology Division, Oak Ridge National Laboratory

Oak Ridge, Tennessee 37830

This paper deals with two interrelated ideas: (a) the relation of cell division to radiation injury and death, and (b) the consequent usefulness of certain irradiated systems for studying the role of cell division in development.

SEPARATION OF "GENETIC" AND "PHYSIOLOGICAL" ASPECTS OF RADIATION-INDUCED LETHALITY

Doses of ionizing radiation that are considered high for producing injury and lethality impart very little energy to the tissues irradiated. For example, a dose of 500 rad of X- or gamma-rays is lethal for man. If this dose were dissipated and retained completely as heat, it would produce a temperature rise of only about 0.001°C. Thus, in terms of total energy imparted to tissue by ionizing radiations, we might not expect *a priori* that even lethal doses would have significant effects on the metabolic machinery at the organizational level usually studied by plant physiologists. The explanation for the great effectiveness of ionizing radiations lies in molecular biophysics and cytogenetics. Radiation cytogeneticists are familiar with chromosome breakage caused by the localized concentration of ionizations[20] within a small region of the chromosome.[21] For radiation-induced damage and cell lethality, the main importance of these chromosomal aberrations lies in the production of nuclear imbalance, which results in deficiencies in large numbers of genes. Nuclear imbalance cannot occur, however, unless the nucleus divides after irradiation. With increasing radiation dose, the extent of chromosome breakage and rearrangement increases. Also with increasing dose the mitotic frequency decreases. After receiving very high doses of gamma-rays, certain seeds germinate with little or no mitosis and cytokinesis; the germinated seeds are called "gamma-plantlets." Despite extensive chromosome breakage and rearrangement, the absence of mitosis prevents the development of nuclear imbalance in gamma-plantlets. The interaction of chromosome breakage and mitosis in producing nuclear imbalance in germinating corn seeds exposed to doses lower than those needed to produce gamma-plantlets was reported in 1956 by Schwartz and Bay.[18] This has been extended to other cereals.[3]

The mechanism of radiation-induced mitotic inhibition is unknown. The mitotic inhibition may or may not be accompanied by complete inhibition of DNA synthesis.[5,6] Absence of DNA synthesis is from three lines of evidence; (a) lack of increase in gross DNA content, as shown by chemical extraction and determination,[19] (b) absence of incorporation of radioactive thymidine into DNA despite the presence of thymidine kinase,[5,6,19] and (c) absence of increase in DNA per nucleus as shown by Feulgen microspectrophotometry.[6] Effective doses, given to dry seeds before sowing, are 500-800 krad for wheat and 1.3 Mrad for lettuce. Radiation-induced inhibition of DNA synthesis and mitosis in gamma-plantlets indicates genetic lethality, in that the reproduction of cells and of the whole organism fails to occur. However, the gamma-plantlets do have many normal features, indicating that the genetic lethality is not accompanied by physiological lethality.

Many of the physiologically normal aspects of gamma-plantlets have been reviewed elsewhere.[3] Some more recent findings are concerned with the normal time course and the chemical regulation of senescence,[2] sequential growth response to growth substances,[17] the relative biological effectiveness (RBE) of neutron irradiation of <u>seedlings</u> apart from cell division,[13] growth in the presence of 5-fluorodeoxyuridine at concentrations that prevent DNA synthesis,[10] development of pyrimidine deoxynucleoside kinase activities,[19] and formation of an initial protrusion that is characteristic of initiation of a new leaf primordium.[1]

"GAMMA-PLANTLETS": USE OF RADIATION-INDUCED MITOTIC INHIBITION IN STUDIES OF DEVELOPMENT

The many uses of gamma-plantlets in studies of various aspects of plant development have been reviewed elsewhere.[3] More recent applications are the demonstration that the initial protrusion of a leaf primordium can form without concurrent, periclinal cell divisions;[1] separation of the onset of thymidine kinase activity from DNA synthesis;[19] testing the specificity of chemical inhibitors of DNA synthesis;[12] and further findings concerning the actions of growth regulators.[17] Of course the usefulness of gamma-plantlets in studies of this type depends on the extent to which the radiation effects are initially specific for preventing DNA synthesis and cell division.

ON THE "SPECIFICITY" OF THE RADIATION EFFECT USED IN MAKING WHEAT GAMMA-PLANTLETS

The compound 5-fluorodeoxyuridine (FdUrd) inhibits DNA synthesis but has little or no inhibitory action on the growth of wheat seedlings, apart from its inhibition of DNA synthesis and mitosis.[12] The growth of wheat seedlings with normal DNA synthesis and mitosis can be inhibited greatly by FdUrd concentrations as low as 0.0001 M. In contrast, the growth of gamma-plantlet seedlings is unaffected by FdUrd concentrations as high as 0.1 M. Comparison of the growth curves of both gamma-plantlet and unirradiated (control) seedlings treated with various concentrations of FdUrd leads to the conclusion that the radiation used to produce gamma-plantlets has little or no inhibitory effect on growth by mechanisms other than the inhibition of DNA synthesis and mitosis.[10] Similar experiments with the mitotic inhibitor maleic hydrazide lead to the same conclusion.[15] Two additional indications that cell growth in gamma-plantlets is inhibited largely as a result of mitotic inhibition are: (a) in embryo culture the initial growth rates of gamma-plantlet leaves growing without cell division are the same as those of unirradiated leaves growing with many cell divisions (see Fig. 2 of Ref. 16); and (b) among those structures (e.g. coleorhiza, epiblast, root hairs) that show no cell division in unirradiated controls, the final size attained in gamma-plantlets is the same as in the controls.[5] The specificity of the radiation effect in this system is comparable to that of FdUrd and much greater than the specificities of other chemicals—e.g. hydroxyurea and phenethyl alcohol—that have been used as inhibitors of DNA synthesis.[10,12] Consequently, we conclude that after irradiation which prevents cell division during germination of wheat, the decreased growth observed relative to unirradiated seedlings can be attributed largely to consequences of the mitotic inhibition and not to other effects of the gamma-radiation.

Despite the surprising degree to which the radiation effect in gamma-plantlets seems to be specific for inhibiting DNA synthesis and cell division, there are still indications of lack of complete specificity. Among these are the loss of sensitivity of leaf growth in gamma-plantlet embryo cultrues to galactose in the presence of optimal concentrations of glucose[16] and the induction of physiological dormancy in lettuce.[8] A possible delay in onset of growth in gamma-plantlets[5] may be an endosperm effect, since it is not observed in embryo cultures.[16]

CELL DIVISION AND SUBLETHAL RADIATION INJURY

Sublethal radiation can induce chromosomal aberrations and nuclear imbalance may result from subsequent division of cells. In the absence of such nuclear imbalance, sublethal radiation effects can also be studied by seedling irradiation of gamma-plantlets,[7,13] which provide a simplified system of studying radiation effects on cell growth uncomplicated by nuclear imbalance and other consequences of cell division. The apical section of normal wheat leaves provides an even simpler system, in which both cell growth and cell division are absent. In studies of three varieties of wheat, gamma-radiation doses as high as 100-400 krad were found to be completely without detectable effect on senescence as measured by rate of chlorophyll loss.[14] By contrast, senescence in these systems is very strikingly regulated, both positively and negatively, by treatment with low concentrations of various chemicals, by temperature, or by illumination with low intensities of light.[2,4,11,14] Thus, systems that are easily regulated by physical and chemical treatments may be unaffected by remarkably high doses of ionizing radiations, provided there is an absence of the typical radiation effects of (a) nuclear imbalance resulting from division of cells with chromosomal aberrations, (b) radiation-induced mitotic inhibition, (c) radiation-induced growth inhibition, and (d) amplification of somatic mutations by cell division.

ESTABLISHING CRITERIA FOR PHYSIOLOGICAL LETHALITY APPLICABLE TO NONDIVIDING SYSTEMS

In the experiments with wheat leaf apices described in the preceding section, we could find no gamma-radiation doses that accelerated senescence without directly killing the leaf tissue.[14] Lethal doses ranged from 250 to 800 krad. Acceleration of senescence in green tissue is indicated by an acceleration of the slow loss of chlorophyll, whether in light or in darkness, relative to controls that have received no ionizing irradiation. By contrast, death is indicated by rapid chlorophll loss in the light and inhibition of chlorophyll loss in the dark.[14] These criteria of lethality are accompanied by the following additional criteria: loss of a temperature optimum for chlorphyll loss in darkness, loss of sensitivity to growth regulators, and incapacity to perform photosynthesis.[14,22] Therefore, in the apical cereal-leaf system, in which there is neither cell division nor growth, direct cell killing can be experimentally separated from acceleration of senescence.

"LAW OF BERGONIE AND TRIBONDEAU"

Our considerations so far have dealt with several aspects of the relation between cell division and radiation injury. The greater effectiveness of ionizing radiation on dividing than on nondividing cells is encompassed in the "Law of Bergonie and Tribondeau." This "law," however, is less a precise scientific law subject to experimental test than a set of generalized views: (a) Radiation effects are more easily seen in dividing then in nondividing cells. (b) Dividing cells have greater intrinsic radiosensitivity. (c) After irradiation, cell divisions increase damage.

The first statement is generally true but essentially trivial since some of the most characteristic cytological radiation effects --mitotic inhibition, gross genic imbalance, and multipolar spindles --could not possibly occur in cells not undergoing mitosis. However, other effects, such as chromosome breakage or somatic mutation, can occur in both dividing and nondividing cells but can be detected only in dividing cells.

The second statement is contradicted by experiments with homoblastic leaf development in tobacco. The rate of cell division decreases as tobacco leaves grow, and older leaves grow without any cell divisions. Disks can therefore be cut from leaves at various developmental stages, undergoing different degrees of cell division, and these can be used for studies of a single biological effect both in dividing and in nondividing tissues of the same morphological and physiological cell type. In the chlorenchyma, induced susceptibility to photodestruction of chloroplasts has been used to measure sensitivity to gamma-radiation as a function of the rate of cell division. Since the criteria used involve death very soon after high doses of radiation, postirradiation cell division had little or no effect on the radiosensitivity that was studied. In this system radiosensitivities were approximately the same, irrespective of the extent of cell division occurring at the time of irradiation.[9]

The third statement, however, is the most relevant. It is a valid generalization which we have dealt with already in connection with the role played by postirradiation cell division in producing nuclear imbalance. Postirradiation cell division also amplifies somatic mutations by perpetuating cell lineages from individual, mutated cells. Thus, extensive cell division after irradiation will result in further damage.

CELL DIVISION AND RBE

From our considerations up to this point we can conclude that cell division plays an important role in the production of physiological radiation injury. A further indication of the importance of cell division in the expression of radiation injury comes from studies comparing the RBE of different radiations in dividing and nondividing systems. Compared with gamma-rays, the RBE of fast neutrons for inhibition of wheat growth is approximately 6-7, provided many cell divisions are required for the final extent of growth observed after irradiation of seedlings. When postirradiation cell divisions are not required for the effect observed, the RBE is only 2-3. This lower RBE is observed (a) in seedlings irradiated and studied within a few days after irradiation, when the growth occurs with relatively few cell divisions; (b) in gamma-plantlets (produced as usual by irradiation of dry grain with gamma-rays) that are irradiated as seedlings with neutrons or gamma-rays and later measured after final cessation of growth without cell division; and (c) in apical wheat leaf sections observed for lethality, which undergo neither cell division nor cell growth.[13] These findings on the influence of cell division on RBE provide an alternative to previous generalizations concerning the dependence of RBE on either the dose administered or the stage of plant development scored.[13] Our experiments point to a nonmitotic (or "physiological") component of the RBE of fast neutrons, with a value of 2-1/2 and a mitotic (or "cytogenetic") component, with a value much greater than 7. Our findings thus indicate a qualitative, not merely quantitative, importance of cell division in the expression of radiation injury.

CONCLUSIONS

Much of the growth inhibition produced by ionizing radiation is a consequence of mitotic inhibition, although systems are available for studying inhibition of cell growth without division. Radiation seems to promote senescence by means of nuclear imbalance and growth inhibition but not significantly by accelerated senescence in absence of growth, cell division, and nuclear imbalance. The role of cell division in radiation injury is not to increase intrinsic radiosensitivity but to amplify radiation-induced damage by producing nuclear imbalance as a result of chromosome breakage and perhaps also perpetuating somatic mutations in the progeny of mutated cells. Another indication of the qualitative importance of cell division in the production of radiation injury comes from findings that, for growth inhibition after irradiation with neutrons, much of the high RBE results from a component requiring cell division. In the absence of the many deleterious consequences of postirradiation cell division, seedlings showing complete genetic lethality can be surprisingly normal physiologically. This last consideration underlies the use of radiation-induced mitotic inhibition in certain systems for studying the relation of cell division to morphogenetic, physiological, and biochemical aspects of plant development.

REFERENCES

1. Foard, D. E. (1971) The initial protrusion of a leaf primordium can form without concurrent periclinal cell divisions. Can. J. Bot. (in press).

2. Foard, D. E. and Haber, A. H. (1970) Physiologically normal senescence in seedlings grown without cell division after massive gamma-irradiation of seeds. Radiat. Res. 42: 372-380.

3. Haber, A. H. (1968) Ionizing radiations as research tools. Ann. Rev. Plant Physiol. 19: 463-489.

4. Haber, A. H. (1971) Chloroplast senescence in nongrowing tissue in darkness: effects of temperature and age. Exp. Gerontol. 6: 179-185.

5. Haber, A. H. and Foard, D. E. (1964) Further studies of gamma-irradiated wheat and their relevance to use of mitotic inhibition for developmental studies. Amer. J. Bot. 51: 151-159.

6. Haber, A. H., Foard, D. E. and Perdue, S. W. (1969) Actions of gibberellic and abscisic acids on lettuce seed germination without actions on nuclear DNA synthesis. Plant Physiol. 44: 463-467.

7. Haber, A. H., Foard, D. E. and Triplett, L. L. (1969) Seedling irradiation of 'gamma-plantlets': demonstration and explanation of a leaf growth stimulation. Radiat. Bot. 9: 473-479.

8. Haber, A. H. and Luippold, H. J. (1959) Dormancy resulting from gamma-irradiation of lettuce seed. Int. J. Radiat. Biol. 1: 317-327.

9. Haber, A. H. and Rothstein, B. E. (1969) Radiosensitivity and rate of cell division: "Law of Bergonie and Tribondeau". Science 163: 1338-1339.

10. Haber, A. H. and Schwarz, O. J. (1970) Actions of FUdR on "gamma-plantlets" and unirradiated plants. Plant Physiol. 46 (suppl.): 31.

11. Haber, A. H., Thompson, P. J., Walne, P. L. and Triplett, L. L. (1969) Nonphotosynthetic retardation of chloroplast senescence by light. Plant Physiol. 44: 1619-1628.

12. Haber, A. H. and Triplett, L. L. (1970) Testing the specificity of inhibitors of DNA synthesis for growth studies. ASB Bull. 17: 45.

13. Haber, A. H., Triplett, L. L., Haywood, F. F. and Ward, D. R. (1970) RBE of fast neutrons in growth inhibition: Evidence for a component requiring cell division. Radiat. Res. 44: 446-455.

14. Haber, A. H. and Walne, P. L. (1968) Actions of acute gamma radiation on excised wheat leaf tissue. I. Radioresistance of senescence and establishing criteria of death from studies of chlorophyll loss. Radiat. Bot. 8: 389-397.

15. Haber, A. H. and White, J. D. (1960) Action of maleic hydrazide on dormancy, cell division, and cell expansion. Plant Physiol. 35: 495-499.

16. Long, T. J. and Haber, A. H. (1965) Growth of embryos on synthetic media after excision from gamma-irradiated (525 kr) and unirradiated wheat grains. Radiat. Bot. 5: 223-231.

17. Rose, R. J. and Adamson, D. (1969) A sequential response to growth substances in coleoptiles from gamma-irradiated wheat. Planta 88: 274-281.

18. Schwartz, D. and Bay, C. E. (1956) Further studies on the reversal in the seedling height dose curve at very high levels of ionizing radiations. Amer. Natur. 90: 323-327.

19. Schwarz, O. J. and Haber, A. H. (1971) The development of pyrimidine nucleoside kinase activities in "gamma-plantlets". Proc. Assoc. Southern Agr. Workers, 68th. Ann. Conv., p. 201.

20. Setlow, R. B. and Pollard, E. C. (1962) Molecular Biophysics. Addison-Wesley, Reading, Mass. 545 pp.

21. Swanson, C. P. (1957) Cytology and Cytogenetics. Prentice-Hall, Inc., Englewood Cliffs, New Jersey, 596 pp.

22. Walne, P. L. and Haber, A. H. (1968) Actions of acute gamma radiation on excised wheat leaf tissue. II. Immediate lethal dose effects on chloroplast ultrastructure and subsequent re-retardation of normal disintegration in darkness. Radiat. Bot. 8: 399-406.

*Research sponsored by the U.S. Atomic Energy Commission under contract with the Union Carbide Corporation.

-DISCUSSION-

DAVIDSON - It should not be assumed that FUdR has effects only on DNA synthesis. In *Vicia* roots it blocks some cells in G_2 and in normal and irradiated barley coleoptiles it inhibits growth.

HABER - In the seedlings we found no non-specific effect but when we treated dry seed we found a non-specific effect. The non-specific inhibition of FUdR can be reversed by thymidine. Using labeled FUdR it appears that its uptake is inhibited by thymidine.

KONZAK - Have you measured protein synthesis in this material?

HABER - Gamma-plantlets grown in embryo culture show a net increase in protein accomplished at the expense of nitrate in the medium. Various new isozymes of enzymes initially present in the plant also occur as demonstrated by acrylamide gel electrophoresis.

KONZAK - How does it compare with control?

HABER - It is about the same.

KONZAK - At the same developmental stage?

HABER - Yes.

CONGER, B. - Gamma-plantlets have been reported for both maize and wheat; have they been described for dicotyledenous species?

HABER - The gamma-plantlet system doesn't work as well with dicotyledons.

CONGER, B. - Does your explanation of your RBE results still hold for the very high RBE's that have been obtained in dry seed systems?

HABER - The generalization I am suggesting is that when the consequences of cell division are important the RBE is high. When cell division is not involved the RBE is relatively low.

SHOOT DRY WEIGHT OF THE SOYBEAN SEEDLING FOLLOWING GAMMA IRRADIATION: EFFECTS OF EXPOSURE, EXPOSURE RATE, AND SPLIT EXPOSURE*

M. J. Constantin and D. D. Killion

Agricultural Research Laboratory of the University of Tennessee, Oak Ridge, Tennessee**

INTRODUCTION

The exposure of organisms to ionizing radiation induces a series of varied effects that collectively can lead to growth reduction and/or lethality. A number of biological, environmental, and radiological factors, however, influence the extent of damage induced by an exposure to ionizing radiation.[15] Results of research with a number of biological test systems show that more damage is induced per unit of radiation delivered at relatively high rates than at low rates, i.e., there is an exposure-rate effect.[5] The split-exposure technique is commonly used to study exposure-rate effects in organisms because the phenomenon is generally interpreted on the basis of repair of induced damage.[2,5,8,11]

Although much of the information pertinent to exposure-rate effects relates to micro-organisms and animals, the phenomenon has been observed in several plant species.[1,4,7,8,9,10,12,15] We are conducting a series of experiments on growth of the soybean seedling as affected by gamma radiation exposure, exposure rate, and split exposure. This report will summarize the results of experiments completed to date; some of these data have been published previously.

MATERIALS AND METHODS

Procedures generally utilized in studies of radiation-induced growth reduction in the soybean seedling have been published.[4] Briefly, they consist of growing seedlings of <u>Glycine max</u> (L.) Merrill 'Hill' in a medium of peat:perlite:sand (1:1:1 volume) in containers under conditions of 28-29° C temperature, 50% relative humidity, and 1000-1500 ft-candles of mixed incandescent and fluorescent light for 16 hours daily. Water and nutrients were supplied as needed to assure adequate growth and development of the seedlings. At 25-30 days after irradiation the shoot growth above the unifoliolate leaf node was harvested, oven-dried at 65° C for 96 hours, and weighed. Growth from axillary buds at the cotyledonary and unifoliolate leaf nodes was either excluded or harvested separately.

Irradiation of seedlings at 2-4 days after emergence, when the unifoliolate leaves were unfolded, was done indoors at the Variable Gamma Dose-Rate Facility of the Agricultural Research Laboratory of the University of Tennessee.[3] Conditions of temperature, humidity, and light intensity and duration in the source building approximated those in the environment-controlled chamber. The design of an experiment to determine the effect of varying the initial exposure is shown in Table I. Exposure rate was 50 R/min for both the initial and second exposure. The radiation-free period between exposures was held constant at 120 min.

TABLE I. DESIGN OF AN EXPERIMENT TO DETERMINE THE EFFECT OF VARIED INITIAL EXPOSURES ON SHOOT DRY WEIGHT OF THE SOYBEAN SEEDLING

Initial	Second exposure, R							
exposure, R	0	500	1000	1500	2000	2500	3000	3500
0	1*					2	3	4
500						5	6	7
1000				8	9	10	11	
1500			12	13	14	15		
2000		16	17	18	19			
2500		20	21	22				

*Treatment number; each treatment was replicated 10 times.

Treatments were selected on the basis of prior results to minimize the number of experimental units. Two seedlings were grown per container per treatment located at random within each of 10 replications. Plants of each treatment in a series subjected to the same initial exposure were irradiated simultaneously. For example, treatments 5, 6, and 7 were irradiated for 10 min (initial exposure), given a 120-min radiation-free period, and irradiated for the respective second exposure. Therefore, plants of treatment 6 received a 1-2 min radiation-free period during the second exposure when plants of treatment 5 were removed from the source of radiation. Likewise, plants of treatment 7 received two such brief radiation-free periods during the second exposure.

RESULTS AND DISCUSSION

A brief description of the soybean shoot apical meristem will provide the histological characteristics of the system being studied. Miksche[13] studied the developmental vegetative morphology of the shoot apical meristem of the soybean variety 'Hawkeye'. Mitotic figures were observed in the apical zone 40-48 hours after the initiation of germination. The third trifoliolate leaf was formed by 130 hours (5.5 days), and subsequent leaves were initiated at a two-day interval. A histological study of 'Hill' soybean seedlings indicated a pattern of development similar to that described above. The shoot apical meristem of 'Hill' has a biseriate tunica, whereas Miksche[13] observed a uniseriate tunica in 'Hawkeye'. Thus the system being irradiated and studied is comprised of organs, tissues, and a population of asynchronous cells varying in their degree of differentiation and mitotic cycle time. At least three trifoliolate leaves in varying stages of development are present in the shoot apical bud at the time of irradiation. The average nuclear volume of tunica cells is $120 \pm 4\ \mu^3$ for seedlings at 2 days after emergence; the average interphase chromosome volume is $3\ \mu^3$. A value of $4.1\ \mu^3$ was listed as the average interphase chromosome volume of the soybean variety 'Hawkeye' and an LD_{50} of 18.9 ± 5.4 kR was predicted[14] for a 36-hour fallout decay simulated exposure.

A typical exposure response curve for the Hill soybean seedling is shown in Figure 1. Data were taken from an experiment in which the exposure rate was 50 R/min and the seedlings were irradiated continuously. The response curve has a shoulder that extends through approximately 2500 R. The mean shoot dry weight for the 2500 R population appears to be abnormally high in comparison to results of other experiments. Although this value is included, it does not affect appreciably the D_{50} (exposure that reduced shoot dry weight to 50% of the control, i.e., non-irradiated seedlings). Increasing the exposure from 2500 to 3500 R caused a steep decrease in shoot dry weight. Seedlings exposed to 4000 R or more expanded trifoliolate leaves that were developed in the shoot apical bud at the time of irradiation. Those leaves were diminutive, however, and no new leaves emerged. The seedlings remained green and turgid for the duration of the experiment.

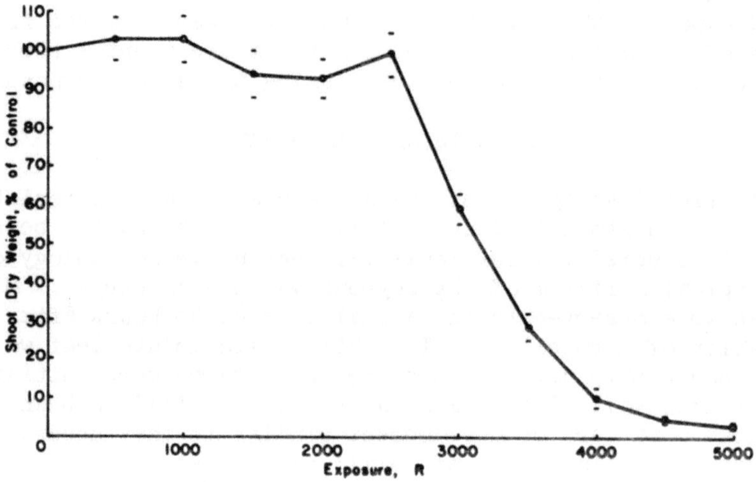

Figure 1. Gamma radiation response curve for shoot dry weight of the Hill soybean seedling.

D_{50} for shoot dry weight was 3100 ± 100 R. Mean shoot dry weight of control seedlings was 1.07 ± 0.15 g. The D_{50} value varies somewhat from one experiment to another although conditions are maintained as similar as possible. A certain amount of variation is to be expected on the basis of the numerous factors that can influence the radiation response of organisms.[15]

In a preliminary experiment, seedlings were exposed to 4000 R at rates of from approximately 3 through 400 R/min.[4] Shoot dry weight was unaffected by exposure rate from 50 through 400 R/min indicating that at 4000 R the Hill soybean shoot apical meristem had incurred an irreparable level of injury at these relatively high exposure rates. On the other hand, shoot dry weight decreased from approximately 75% at approximately 3 to < 10% at 50 R/min. Exposure response curves for seedlings exposed at either 10 or 50 R/min[4] are shown in Figure 2. D_{50}'s were approximately 5000 and 3000 R for seedlings exposed at 10 and 50 R/min, respectively. The exposure-rate effect was at a minimum at the lowest and highest exposures and it was at a maximum at the intermediate exposure, 4000 R. This type of response is probably caused by a limitation of the test system analogous to that observed in mice and discussed by Krebs and Leong.[11] Similar observations have been reported by

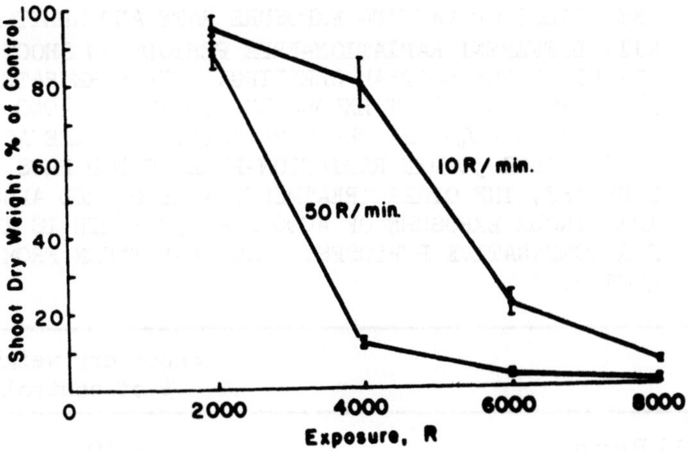

Figure 2. The effect of varying exposure rate (10 versus 50 R/min) on the gamma radiation response of the Hill soybean seedling (redrawn from reference 4).

Bottino and Sparrow[1] for barley, Hall and Bedford[8] for *Vicia*, and McCrory and Grun[12] for *Solanum* hybrids. Experimental results indicate a significant first-order interaction between exposure and exposure rate, and a second-order interaction involving exposure, exposure rate, and developmental stage of the plant at the time of irradiation.[4,9,10]

Further evidence of an exposure-rate effect was obtained in another experiment; data are shown in Table II. If either rate was decreased or the total exposure was split equally, shoot dry weight was greater than that following 4000 R at 50 R/min. However, a decrease in rate from 50 to 20 R/min caused a six-fold increase and a split exposure with a 120-min radiation-free period caused a four-fold increase. No difference in shoot dry weight was observed between 4000 R (36.4 R/min continuous) and 4000 R (50 R/min split) with a 30-min radiation-free period between exposures. In contrast, shoot dry weight was greater following 4000 R (20 R/min continuous) than 4000 R (50 R/min split) administered with a radiation-free period of 120 min between exposures.

In three different experiments, various combinations of initial and second exposures were used to equal 4000 R accumulated with a 120-min radiation-free period and an exposure rate of 50 R/min. Shoot dry weight of the seedlings ranged from approximately 10 to 39% of control with an average value of 20% (data not shown). A

TABLE II. THE EFFECT OF VARYING EXPOSURE RATE AND SPLIT EXPOSURE WITH DIFFERENT RADIATION-FREE PERIODS ON SHOOT DRY WEIGHT OF THE SOYBEAN SEEDLING. EXPOSURE RATE WAS 50 R/MIN FOR THE SPLIT EXPOSURES, THEREFORE 4000 R at 36.4 R/MIN REQUIRED THE SAME TREATMENT TIME AS DID 2000 R PLUS 30-MIN RADIATION-FREE PERIOD PLUS 2000 R. LIKEWISE, THE OTHER TREATMENTS WERE PAIRED ALSO, AND THE SINGLE EXPOSURE OF 4000 R AT 50 R/MIN IS INCLUDED FOR COMPARATIVE PURPOSES. DATA ARE TAKEN FROM REFERENCE 4.

Treatment	Shoot dry weight, % of control
4000 R @ 50 R/min	$\leq$ 10
4000 R @ 36.4 R/min	15 ± 1[1]
2000 R + 30 Min + 2000 R	16 ± 2
4000 R @ 28.6 R/min	43 ± 4
2000 R + 60 Min + 2000 R	34 ± 3
4000 R @ 20 R/min	60 ± 3
2000 R + 120 Min + 2000 R	39 ± 3

[1] Mean ± standard error

4000 R continuous exposure at 50 R/min reduced shoot dry weight to $\leq$ 10% of control in repeated experiments. When seedlings received an initial exposure of 2000 R, a 120-min radiation-free period, plus a second exposure of 0, 500, 1000, 1500, 2000, 2500, or 3000 R at 50 R/min, the D_{50} for shoot dry weight was 1700 ± 50 R (Figure 3). The initial exposure of 2000 R plus the second exposure of 1700 R equals 3700 R to reduce shoot dry weight to 50% of control. This represents an increase in the D_{50} value of approximately 600 R when compared to a continuous exposure at 50 R/min.

At this time we decided to determine whether the phenomenon would show a response to varying initial exposures, i.e., would there be an optimal first exposure. The experimental design is shown in Table I. The influence of varying the initial exposure on D_{50} for shoot dry weight is shown in Table III. D_{50} was

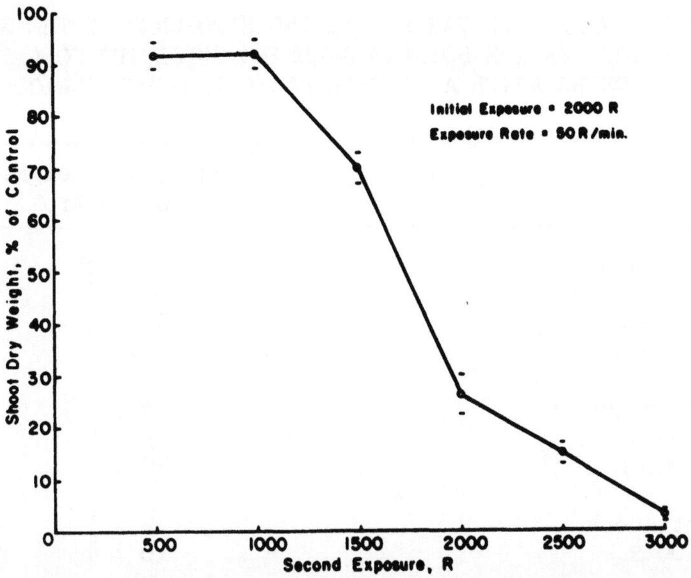

Figure 3. Gamma radiation response curve for shoot dry weight of soybean seedlings subjected to an initial exposure of 2000 R with a 120-min radiation-free period.

TABLE III. CHANGES IN D_{50} FOR SHOOT DRY WEIGHT OF THE SOYBEAN SEEDLING INDUCED BY SPLIT EXPOSURES WITH VARIED INITIAL EXPOSURES. THE RADIATION-FREE PERIOD WAS 120 MIN.

Initial exposure, R	D_{50}, R	Change in D_{50}, R
0	2750 ± 150	---
500	3700 ± 30	950
1000	3550 ± 70	800
1500	3470 ± 100	720
2000	3400 ± 80	650
2500	< 3000*	---*

*An initial exposure of 2500 R plus a second exposure of 500 R, the lowest used, caused a reduction in shoot dry weight greater than 50% of control.

TABLE IV. THE EFFECT OF VARIED INITIAL EXPOSURES ON THE SHOOT DRY WEIGHT OF THE SOYBEAN SEEDLING SUBJECTED TO A SECOND EXPOSURE AFTER A 120-MIN RADIATION-FREE PERIOD.

Treatment	Shoot dry weight, % of control
2500 R continuous	67 ± 3
500 R + 2500 R	87 ± 6
1000 R + 2500 R	54 ± 6
1500 R + 2500 R	20 ± 5
3000 R continuous	3 ± 1
500 R + 3000 R	73 ± 4
1000 R + 3000 R	20 ± 5
3500 R continuous	5 ± 3
500 R + 3500 R	16 ± 4

2750 ± 150 R for continuous irradiation. This value is considerably lower than that observed in previous experiments. Mean shoot dry weight of control seedlings was 1.05 ± 0.08 g. A probable explanation is that we switched to a supply of new seeds obtained from a commercial seed firm.* The seeds used for earlier experiments were obtained from a supply of seeds used for variety testing.** The stability of the radiation response of seedlings from the new supply of seeds has not been verified.

An initial exposure of 500 R increased the D_{50} by 950 R. The increase in D_{50} was greater than the initial exposure which is inconsistent with the concept of a repair phenomenon. Each additional increment in the initial exposure tended to decrease the D_{50}. Accordingly, an initial exposure of 2500 R followed by a second exposure of 500 R, the lowest used, reduced shoot dry weight to less than 50% of control. Although the experiment was designed to study

*Courtesy of Dobson-Hicks Supplies, Nashville, Tennessee.

**Courtesy of C. R. Graves, Department of Agronomy, University of Tennessee, Knoxville.

the change in D_{50}, the data can be considered as shown in Table IV. Admittedly, the data are limited, however, a 2500 R continuous exposure reduced shoot dry weight to 67 ± 3% and 2500 R after an initial exposure of 500 R reduced shoot dry weight to 87 ± 6% of control. Higher initial exposures did not induce an increase in shoot dry weight at 2500 R, but did so at 3000 R. Shoot dry weight was 3 ± 1% of control following a continuous exposure of 3000 R and it was increased to 73 ± 4% following an initial exposure of 500 R. An initial exposure of 1000 R caused an increase in shoot dry weight to 20 ± 5% of control.

The term "repair" in reference to radiobiological usage refers to a reversal of the injury process prior to the expression of the effect.[16] In the soybean shoot dry weight system, there was an increase in performance when exposure rate was decreased below 50 R/min (see Fig. 2 and Table II). Also, in split-exposure experiments, an initial exposure of approximately two-thirds of the D_{50} caused an increase in shoot dry weight when compared to that of seedlings exposed continuously (see Table II). Such a response is consistent with the phenomenon of repair of radiation-induced injury.

It would appear that the soybean shoot dry weight system also shows a "protective effect" when subjected to various split exposures. The data in Table IV show, for example, that seedlings subjected to an initial exposure of 500 R had a greater shoot dry weight when subsequently exposed to 3000 R than did seedlings exposed to 3000 R without an initial exposure. Also, data in Table III show an increase in D_{50} of 950 R when seedlings were exposed initially to 500 R and given a radiation-free period of 120 min prior to the second exposure. Such a response is consistent with the phenomenon of induced protection. Broertjes[2] observed a similar phenomenon in African Violet petioles exposed to either fast neutrons or gamma radiation. Induced protection was used as an explanation for the phenomenon, however he was unable to determine the basic mechanism involved.

Obviously there are a number of processes that could be operative in the soybean seedling relative to the response observed. Evans[7] has reviewed his research and that of Sax _et al_. relative to repair via chromosomal restitution, and it appears that this type of repair occurs within minutes to hours post-irradiation. Another process is the accumulation of cells in a relatively tolerant phase of the mitotic cycle through mitotic delay induced by the initial exposure.[6] Other processes, such as enzymatic action and auxin balance might be involved, however these are even more speculative than the ones mentioned above.

SUMMARY

Shoot dry weight of the Hill soybean seedling has been used in a number of experiments to study the effects of gamma radiation exposure, exposure rate, and split exposure. Shoot dry weight is influenced by exposure rate such that D_{50} was approximately either 3000 or 5000 R at rates of 50 and 10 R/min, respectively. Seedlings subjected to a split exposure had a greater shoot dry weight than did those exposed continuously. At 4000 R accumulated exposure, seedlings given a 30, 60, or 120-min period between exposures showed an increase in shoot dry weight associated with the duration of the radiation-free period. This response is consistent with the phenomenon of repair of radiation-induced injury.

An initial exposure of 500 R increased the D_{50} value by 950 R in seedlings given a radiation-free period of 120 min. Also, an initial exposure of 500 R reduced shoot dry weight to 73% of control for seedlings subsequently exposed to 3000 R as compared to 3% of control for seedlings exposed to 3000 R continuously, i.e., without an initial exposure. This response is consistent with the phenomenon of induced protection.

The mechanism(s) involved in the soybean seedling has not been investigated.

REFERENCES

1. Bottino, P. J. and Sparrow, A. H. (1970) The effect of exposure time and rate on the survival and yield of lettuce, barley, and wheat. Survival of Food Crops and Livestock in the Event of Nuclear War. Proceedings of Conference, Brookhaven National Laboratory, September 1970, USAEC, DTIE Conference 700909. In press.

2. Broertjes, C. (1970) Dose fractionation studies and radiation-induced protection phenomena in African Violets. Survival of Food Crops and Livestock in the Event of Nuclear War. Proceedings of Conference, Brookhaven National Laboratory, September 1970, USAEC, DTIE Conference 700909. In press.

3. Cheka, J. S., Robinson, E. M., Wade, L. Jr. and Gramly, W. A. (1971) The UT-AEC Agricultural Research Laboratory variable gamma dose-rate facility. Health Phys. 20: 337-340.

4. Constantin, M. J., Killion, D. D. and Siemer, E. G. (1970) Exposure rate effects on soybean plant responses to gamma irradiation. Survival of Food Crops and Livestock in the Event of Nuclear War. Proceedings of Conference, Brookhaven National Laboratory, September 1970, USAEC, DTIE Conference 700909. In press.

5. Dose Rate in Mammalian Radiobiology. (Proceedings of Symposium, Oak Ridge, Tennessee, April 29-May 1, 1968; USAEC, DTIE Conference 680410; Editors: D. G. Brown, R. G. Cragle and T. R. Noonan.)

6. Elkind, M. M., Kamper, C., Moses, W. B. and Sutton-Gilbert, H. (1967) Sublethal-lethal radiation damage and repair in mammalian cells. Recovery and Repair Mechanisms in Radiobiology, Brookhaven Symposia in Biology No. 20, June 5-7, BNL50058(C-51), TID-4500. pp. 134-160.

7. Evans, H. J. (1967) Repair and recovery at chromosome and cellular levels: Similarities and differences. Recovery and Repair Mechanisms in Radiobiology, Brookhaven Symposia in Biology No. 20, June 5-7, BNL50058(C-51), TID-4500. pp. 111-133.

8. Hall, E. J. and Bedford, J. S. (1964) A comparison of the effect of acute and protracted gamma radiation on the growth of seedlings of *Vicia faba*. Part I. Experimental observations. Intern. J. Radiation Biol. 8: 457-474.

9. Killion, D. D., Constantin, M. J. and Siemer, E. G. (1970) Acute gamma irradiation of the soybean plant: Effects of exposure, exposure rate, and developmental stage on survival, height and yield. Radiat. Bot. In press.

10. Killion, D. D. and Constantin, M. J. (1971) Acute gamma irradiation of the wheat plant: Effects of exposure, exposure rate, and developmental stage on survival, height, and grain yield. Radiat. Bot. In press.

11. Krebs, J. S. and Leong, G. F. (1970) Effect of exposure rate on the gastrointestinal LD_{50} of mice exposed to ^{60}Co gamma rays or 250 kVp x-rays. Radiation Res. 42: 601-613.

12. McCrory, G. L. and Grun, P. (1969) Relationship between radiation dose rate and lethality of diploid clones of *Solanum*. Radiat. Bot. 9: 27-32.

13. Miksche, J. P. (1961) Developmental vegetative morphology of *Glycine max*. Agron. J. 53(2): 121-128.

14. Sparrow, A. H. and Puglielli, Leanne (1969) Effects of similated radioactive fallout decay on growth and yield of cabbage, maize, peas and radish. Radiat. Bot. 9: 77-92.

15. Sparrow, A. H., Schwemmer, Susan S. and Bottino, P. J. (1971) The effects of external gamma radiation from radioactive fallout on plants with special references to crop production. Presented at Conference on Survival of Food Crops and Livestock in the Event of Nuclear War. Brookhaven National Laboratory, September 1970, USAEC, DTIE Conference 700909. Radiat. Bot. 11(2): 85-118.

16. Storer, J. B. (1968) Introduction. Dose Rate in Mammalian Radiation Biology (Proceedings of Symposium, Oak Ridge, Tennessee, April 29-May 1; USAEC, DTIE Conference 680410; Editors: D. G. Brown, R. G. Cragle and T. R. Noonan). pp. 1.1-1.4.

*This manuscript is published with the permission of the Dean of the University of Tennessee Agricultural Experiment Station, Knoxville. The research was supported by the Office of Civil Defense and the Atomic Energy Commission.

**Operated by the Tennessee Agricultural Experiment Station for the U.S. Atomic Energy Commission under Contract No. AT-40-1-GEN-242.

-DISCUSSION-

HABER - We found a dose fractionation effect in the growth of the gamma-plantlets (Haber, A., Rad. Bot., 9:473-479, 1969). The effect of radiation was less when the dose was fractionated with an interval of 4 hrs. This would indicate that cell repopulation is not necessary for all such phenomena.

CONGER, A. - Are these plants growing rapidly at the time of irradiation?

CONSTANTIN - No, they were in the lag phase of growth during the first two days (i.e., before they had entered the logarithmic phase).

CONGER, A. - Does 500r given at the first time have the same effect as 500r given at the second time, two hours later?

CONSTANTIN - We have taken the system and irradiated daily from the day of emergence through 16 days and plotted increase in growth; we found no difference in response, at least suggesting that there is no change in sensitivity during this period.

KONZAK - At what point in the development did you measure?

CONSTANTIN - Approximately 25 days after emergence.

KONZAK - How does this relate to the growth increase curve?

CONSTANTIN - Non-irradiated plants are still in their log phase of growth but they are approaching their plateau. Plants that received high exposures have already plateaued.

KONZAK - There may be some differences in the rate of growth of the irradiated and control plants and this could affect your estimate of the effect of radiation on growth.

CONSTANTIN - This is possible. If you plot the growth rate of irradiated and control soybean seedlings during the early period the curves for individual seedlings start to show growth retardation and the extent of retardation is a function of exposure level, as is the point when they plateau.

KONZAK - The only safe place to measure is when the growth has plateaued -- even the controls.

CONSTANTIN - This may be so but we could not take the plants to maturation in the facilities available.

LANGE - May I suggest that it is better to determine the LD_{50} using probit analysis for that also provides a slope and precision estimates; otherwise you are throwing away a large amount of information.

CONSTANTIN - We have done this and we have a family of curves, the slopes of which are essentially parallel. The time at which it plateaus, the level of performance at which it plateaus is dependent upon exposure, dose rate, and on the developmental stage of the plant when it was irradiated. Here we irradiated two to three day seedlings and they are being scored approximately 25 days after irradiation.

BALL - It has been demonstrated several times in a variety of tissue cultures that a pre-treatment at high temperatures enhances the survival of cells when cultures are grown at a later time at prolonged high temperature; and in some cases tissues can be obtained which will grow and flourish at high temperatures. This has not been explained satisfactorily. The cells appear to have a "memory" of their previous history.

CONSTANTIN - The cells have already, as it were, become conditional prior to the second treatment.

KONZAK - I think you ought to check peroxidase levels during the two-hour period. Induction of various enzymes is very common in many systems.

CONSTANTIN - I think it much more likely that we have shifts in cell cycle and populations during this two-hour period giving rise to greater resistance though we don't have any experimental data on this point.

SMITH - There are some very surprising things that happen in split dose experiments with dry seeds if soaking and drying back is involved between fractions. This latter treatment alone confers some resistance to subsequent radiation, so that it is difficult to divorce it from a true fractionation effect. In continually dry seeds fractionation has no effect.

CONSTANTIN - Dose rate and equivalent split dose responses are always more evident in actively growing than in inactive systems.

KONZAK - The soaking between dose fractionation might have leached out protective substances. A water soaking treatment of 15 minutes will remove quite a lot from a seed.

SOME PROBLEMS OF CHROMOSOMAL ABERRATION STUDIES IN MERISTEMS *

JOHN R. K. SAVAGE and MORTON W. MILLER

M.R.C. Radiobiology Unit, Harwell, Didcot, Berks, U.K.

Dept. of Radiation Biology, The University of Rochester

Rochester, New York, U.S.A.

INTRODUCTION

Both dormant and active root meristems are frequently used in the study of radiation cytology as if they were homogeneous cell systems. This assumption is very far from the truth, for heterogeneity is present with regard to cell type, cell stage, cycle and phase times, and the metabolic status of the cells. Moreover, the dormant root is "dormant" when irradiated but "active" when fixed for observation, and the mobilization process introduces additional complications not normally present in roots that have been growing for some time.

It is the purpose of this paper to discuss the consequences of heterogeneity for three aspects of the study of radiation induced chromosomal aberrations: The uniformity of radiosensitivity, the precision with which cells can be located in G_2, and the interpretation of yield-time curves for chromatid aberrations.

THE UNIFORMITY OF RADIOSENSITIVITY

A cell system where all the cells are uniformly radiosensitive has tremendous practical advantages. Firstly, a unique aberration yield for a given dose can be obtained from a single sample of first division metaphases drawn at any fixation time after treatment. Secondly, differential mitotic delay and consequent cell mixing will be absent at each dose, so that the differences in these factors which normally occur between doses will not warp the dose response curve, and interfere with the derivation of meaningful parameters.

Such an ideal situation appears to be approached in the G_1 stage of the developing microspore population of Tradescantia paludosa, a cell system which has been used extensively for the study of radiation-induced chromosome aberrations[2,3,6].

In many dormant root meristems, the majority of the cells are "resting" in the G_1 (or G_0) stage of the mitotic cycle. Accordingly, only chromosome-type aberrations are observed when these cells enter division. It is often assumed, possibly by analogy to the situation in Tradescantia, that cells in G_1 always have a uniform sensitivity, and therefore that it is safe to utilize a single fixation time to assess aberration yield, regardless of the dose or treatment given.

Recently, this assumption has been challenged and shown to be invalid for dormant root meristems of Hordeum vulgare[7] and Vicia faba[4].

Figure 1 shows chromosome exchange yields at various fixation times following single doses of radiation given to the dormant seeds, and it is obvious that the meristem cells do not have a uniform response to radiation. This being so, a single fixation time cannot give a specific aberration yield characteristic of a given dose. More important still, different doses and treatment regimes will, as the result of differential mitotic delay and cell mixing, produce quite different populations of cells, and even multiple sampling will not help when making comparison between treatments. Changes in aberration distribution patterns with fixation time[7] confirm the occurrence of considerable cell mixing.

The reasons for this heterogeneity are not obvious. In both cases cited above, low LET radiations were used. If the dormant cells have a variable oxygen status at the time of radiation, differences in both aberration yield and mitotic delay would be expected[2,5]. One possible test for this would be to repeat experiments of this type with high LET radiation where dependence upon oxygen is very much reduced.

A second fact, which might be of considerable importance is that the cell population at the time of sampling is not in steady state. Indeed evidence from the mitotic index patterns[4,7] suggest a state of considerable flux. This is hardly surprising since the hitherto dormant meristem is now in the process of mobilization for active root growth. All the potential meristematic cells do not

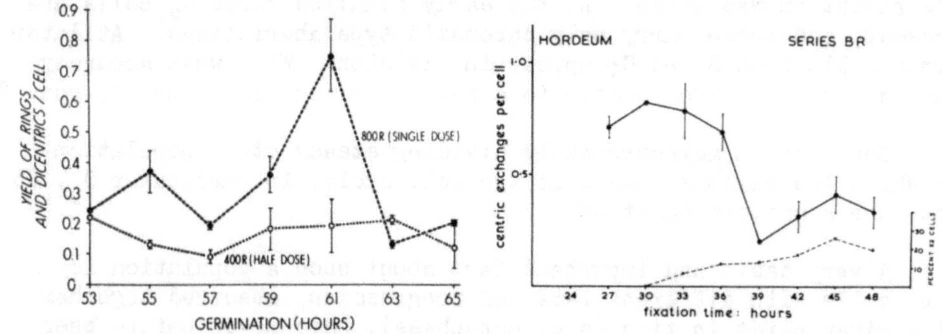

Figure 1. Variation in chromosome exchange yields at various fixation times after single doses of radiation given to dormant seed. Left: Vicia faba, X-rays 400 and 800r (From Reference 4.) Right: Hordeum vulgare, gamma rays 40 Krads. The lower dashed curve is the percentage of recognisable second division (X_2) cells in the sample. (From Reference 7.)

enter division at the same time, i.e. there is no synchronous wave of mitosis. Instead, there is a gradual increase over several hours, and considerable oscillation before a stable M.I. is achieved, by which time many cells are in second and third mitosis. It is a well known fact that in actively growing roots, cells from different parts of a meristem may have widely differing cycle times[1]. In a germinating seed, there are in addition the possibilities that cells may also have different rates of mobilization, or that they are "resting" in quite different regions of the G_1 phase. Add to this the perturbations which radiation introduces into a cell population, and fluctuations in aberration yield with time may not be so surprising.

Whatever the factors responsible, it is clear that we are quite wrong in assuming the root meristem to be a simple cell system.

THE PRECISION WITH WHICH CELLS CAN BE LOCATED IN G_2

Turning now from the dormant meristem to the actively growing root we find an asynchronously dividing cell population which is frequently in a more-or-less steady state, showing a constant mitotic index. If a dose of radiation is given, and serial fixations are commenced immediately afterwards, we can sample cells in metaphase which were at different parts of the cell cycle when the radiation was given. At the early fixation times G_2 cells are present, and these carry only chromatid-type aberrations. At later times cells from S and G_1 appear in division. With what accuracy can we locate a given sample to a particular region of the G_2 phase?

Consider an asynchronously dividing steady state population in which the various phases of the cell cycle, in particular G_2, do not have a uniform duration.

A very basic and important fact about such a population is that cells with different rates of progression, observed together at a given point in time (e.g. metaphase), have never before been together at exactly the same stage of the cell cycle, and will never be so again.

In order to evaluate the effects of variable phase duration we will make two simplifying assumptions. Firstly, the duration of metaphase is very short compared with the minimum G_2 duration, and can be neglected. Secondly, the S/G_2 border has been clearly defined by tritiated thymidine, and neither the endogenous tritium radiation nor any externally applied radiation delays or perturbs the orderly progress of cells towards division.

Let ϕ be the transit time for the compartment G_2 + prophase (G_2P), i.e. the time taken by a cell in passing from the end of S to metaphase. Since ϕ varies from cell to cell in the population, it follows that the cells observed together in a metaphase sample at a particular fixation time must have been in different positions in G_2P at the time of irradiation. It can be shown that the range of positions represented in any sample is dependent upon:

a) The range of ϕ in the population. Assuming, for practical purposes, that a definite minimum and maximum ϕ can be established, this range may be conveniently expressed as the ratio:

$$R_\phi = \frac{\text{maximum } \phi}{\text{minimum } \phi}$$

b) Minimum ϕ

c) The lapse of time between treatment and fixation.

The relative frequency of cells from any particular region of G_2P represented in the sample is dependent upon the form of the ϕ distribution, or a truncated version of it when the fixation time exceeds minimum ϕ.

Figure 2 illustrates these facts for a population where minimum ϕ = 3.0 hours, R_ϕ = 4.0 and the ϕ distribution is log-

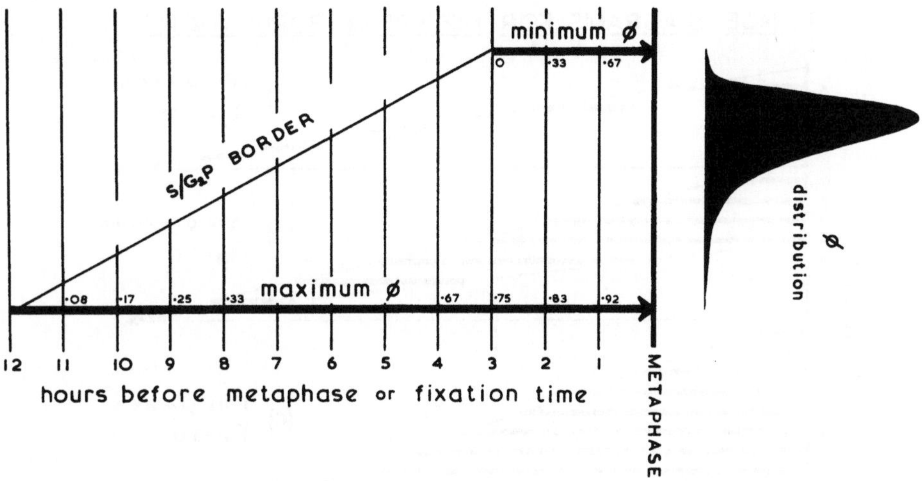

Figure 2. The range of cell positions in G_2P represented in metaphase samples fixed at various times from an ideal population having minimum ϕ = 3.0 hours and R_ϕ = 4.0. The ϕ duration distribution is log-normal, maximum ϕ corresponding to 3.1σ.

normal, maximum ϕ corresponding to +3.1σ. In a steady state population, cells with all values of ϕ will appear together in metaphase. When there is no delay or perturbation, the expected time of arrival in metaphase of a cell irradiated in any particular position in G_2P, will correspond to the fixation time for observing that cell in an unirradiated population. Hence, in the population of Figure 2, a sample of metaphases fixed 1 hour after irradiation will contain cells which were 0.67 of the way through G_2P when irradiated, together with cells which were in all intermediate positions up to 0.92, a range of 25% of G_2P. At the two hour fixation the range of positions represented covers 50% of G_2P (0.33 - 0.83) and at 3 hours the first 75% of G_2P is present in the sample. When, as in this population, the S/G_2P border is clearly defined, the range of G_2P positions increases to a maximum at a time corresponding to minimum ϕ and thereafter declines.

Of course, the relative frequency of cells at the extremes of the range will be low, and the bulk of the sample will correspond to the modal region of the ϕ distribution. However, since ϕ cannot be determined for individual cells observed in metaphase, the whole range will be utilised when computing quantitative values.

A further point to notice is that there is some overlap of ranges, so that cells from the same position in G_2P are present at

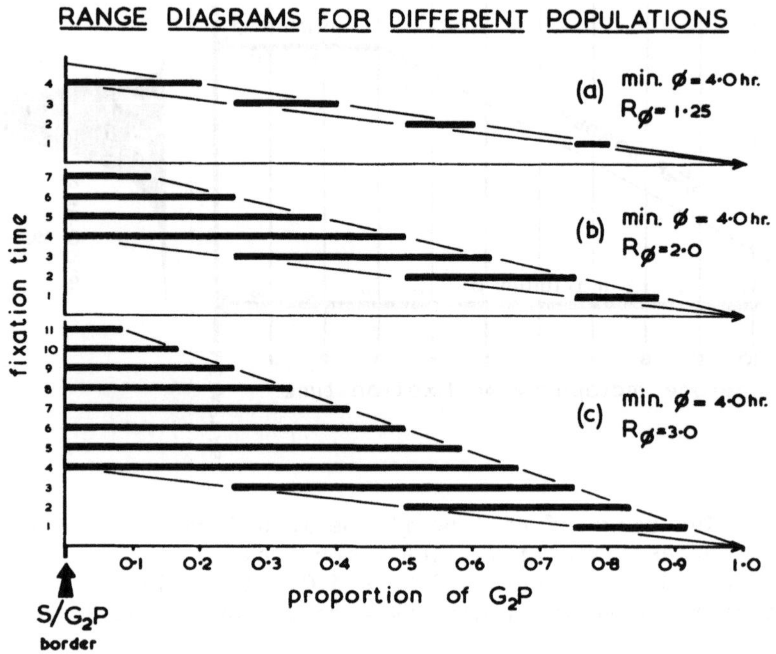

Figure 3. Range diagrams for 3 different populations. The horizontal bars show the proportion of G_2P which will be observed in a metaphase sample at a given fixation time. As the value of R_ϕ increases, ranges at successive fixation times overlap, so that the same region of G_2P is represented at several fixation times.

more than one fixation time, albeit with different frequencies. The degree of range overlap, like the range of G_2P positions present in a sample, is determined by the parameters minimum ϕ and R_ϕ as can be seen by the construction of "range diagrams", examples of which are shown in Figure 3. The ordinate of such diagrams is fixation time (or time before metaphase), whilst the abcissa represents ϕ as a unit vector. The range of G_2P positions, present in a metaphase sample at any fixation time lies between straight lines joining 1.0 on the abcissa to minimum and maximum ϕ on the ordinate.

It can be seen immediately from Fig. 3 that the precision with which we can examine specific regions of G_2P decreases rapidly as R_ϕ increases, because both the range of ϕ present in a sample, and the amount of overlap in successive samples, increases. If our interest centres upon the radiosensitivity of a particular stage or position in G_2P then it is obvious that we can only locate that region in terms of probability, based upon the statistics of the appropriate ϕ distribution for the sample fixation time. This makes quantitative work very difficult. The same argument applies to individual cells; their position in G_2P at the time of irradiation can only be assigned as a probability.

So far we have considered the situation for an ideal population. Neither of the two simplifying assumptions made above hold in reality.

Metaphase has a finite duration, which may itself vary from cell to cell. Moreover, in radiation cytology it is almost a universal practice to add a spindle inhibitor for some hours prior to fixation in order to accumulate metaphases, and facilitate the spreading of the chromosomes. Such treatment lengthens the duration of metaphase, allowing cells to be present in the sample which would normally have been observed only at earlier fixation times. Under these conditions, the range of G_2P positions present in the sample is markedly extended, and, what is more important, the relative frequency of cells from different regions is no longer a simple function of the ϕ distribution.

Radiation, both internal from tritiated thymidine, and also external, certainly perturbs the orderly progress of cells towards division. One effect of this is that the simple relationship between "time before metaphase" and "fixation time", which was assumed for our ideal population, no longer holds. Another overall effect of the perturbation is to change the shape of the ϕ distribution, leading to an increase in minimum ϕ and R_ϕ. This can be seen from the tritiated thymidine labelling index curves following irradiation[8], since the ascending limb of such curves reflects, to some extent, the shape of the ϕ distribution.

Now, as we have shown above, minimum ϕ and R_ϕ are the very parameters which determine the precision with which parts of G_2P can be located. It follows therefore that radiation, in increasing minimum ϕ and R_ϕ reduces this precision, which, even in the control population (with no radiation other than that from the tritiated thymidine), is already rather poor in Vicia, ($R_\phi > 3.0$).

THE INTERPRETATION OF YIELD-TIME CURVES FOR CHROMATID ABERRATIONS

The problem of precision discussed in the previous section has an important bearing upon the quantitative assessment of aberration yield, which is normally expressed as an unweighted arithmetic mean, aberrations per (unlabelled) cell.

Numerous investigators have shown that for a given dose, this mean yield fluctuates throughout G_2P, and this has been taken as an indication of non-uniform radiosensitivity of cells within this compartment. If this is so, can the observed changes in yield be related directly to changes in radiosensitivity throughout G_2P? Furthermore, since there cannot be a unique yield for a given dose, is the observed peak frequency a reliable measure of radiation damage?

We can attempt to answer these questions by a simple extension of the principle of the range diagram. As before, assume an ideal population with the following characteristics:

1) Metaphase duration is short and can be neglected.

2) The S/G_2P boundary is clearly defined with tritiated thymidine so that aberration yields in G_2P (unlabelled) cells only can be determined.

3) There is no cell perturbation resulting from the internal or external radiation.

4) The ϕ distribution is log-normal with a constant $\sigma = 0.35439$. Maximum ϕ is taken at 3.1σ, corresponding to a cumulative area of 0.999.

In order to demonstrate the effects of variation in minimum ϕ and R_ϕ on aberration yield it is necessary to make some assumptions about radiosensitivity throughout G_2P. We are going to assume that a cell is sensitive to radiation (i.e. will carry observable chromatid aberrations) only during 10% of its transit through G_2P. For the remaining 90% it is completely insensitive, no aberrations being produced. In the first instance we will locate this sensitive region between 0.3 and 0.4 of G_2P. All cells within this region at the time of irradiation are uniformly sensitive.

A range diagram for a population where minimum $\phi = 3$ hr. and $R_\phi = 4.0$ is shown in Figure 4, and the boundaries of the sensitive region are indicated. At any fixation time, the frequency of aberration bearing cells corresponds to the area of the ϕ distribution lying within the sensitive region. The arithmetic mean yield of aberrations is therefore directly proportional to this area divided by the total area of G_2P cells, "the relative sensitive area". At fixation times exceeding minimum ϕ, the divisor is the residual area of a truncated ϕ distribution, (stippled in Figure 4), since some of the population is pre-G_2P at these times. These pre-G_2P cells are, of course, in S, so the truncated area corresponds directly to a metaphase labelling index with continuous thymidine treatment.

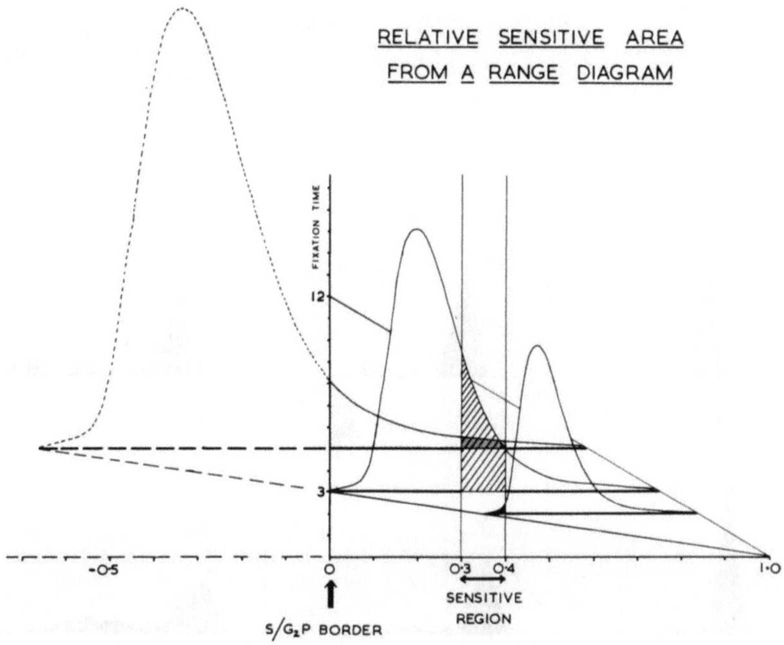

Figure 4. Range diagram for a population where minimum ϕ = 3.0 hr. and R_ϕ = 4.0. Log-normal ϕ distributions are superimposed upon the three ranges shown. The areas of these distributions lying within the sensitive region represent the frequency of aberration bearing cells.

Using discreet, integral fixation times, the relative sensitive area has been computed for nine populations differing in values of minimum ϕ and R_ϕ. The results are shown in Figure 5. It is clear that changes in these parameters markedly affect the shape of the yield-time curve, the position of the peak frequency, and its height (= maximum frequency). Briefly, an increase in R_ϕ lengthens the "tail" i.e. the range of fixation times where aberrations will be observed, without changing the position of the peak. One would not suspect or deduce from these curves that in every case only 10% of G_2P lying between 0.3-0.4 was sensitive.

The great variation peak height is, in part, due to the use of integral fixation times in these ideal populations. There will obviously be a true maximum yield when the mode of the ϕ distribution lies in the sensitive region, and this peak will be

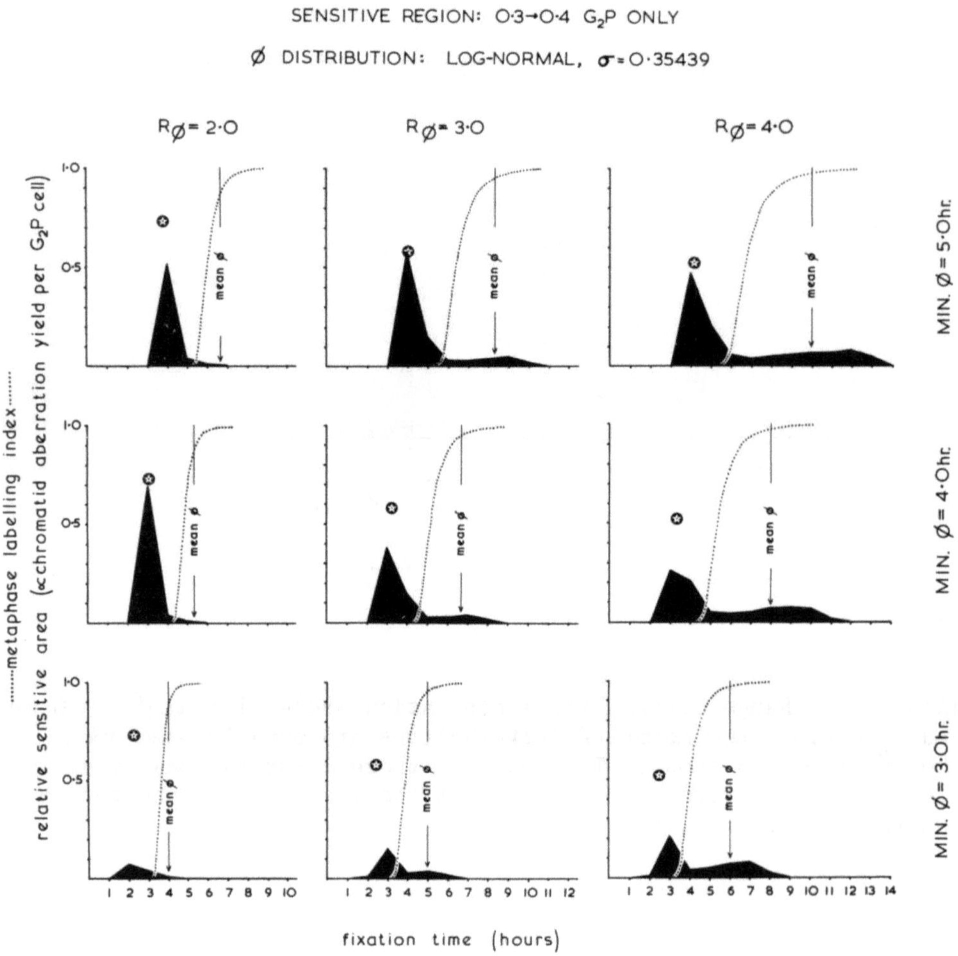

Figure 5. Theoretical yield-time curves of chromatid aberrations in G_2P cells for populations with different values of minimum ϕ and R_ϕ. Metaphase labelling index curves in the presence of continuous tritiated thymidine are also shown.

very sharp. These true peaks and their positions are shown by the stars in Figure 5. We now see that whilst the position of the peak is still dependent upon minimum ϕ, its height is related to R_ϕ. The position of the true peak in relation to the known mean does not indicate a sensitive region between 0.3-0.4 G_2P.

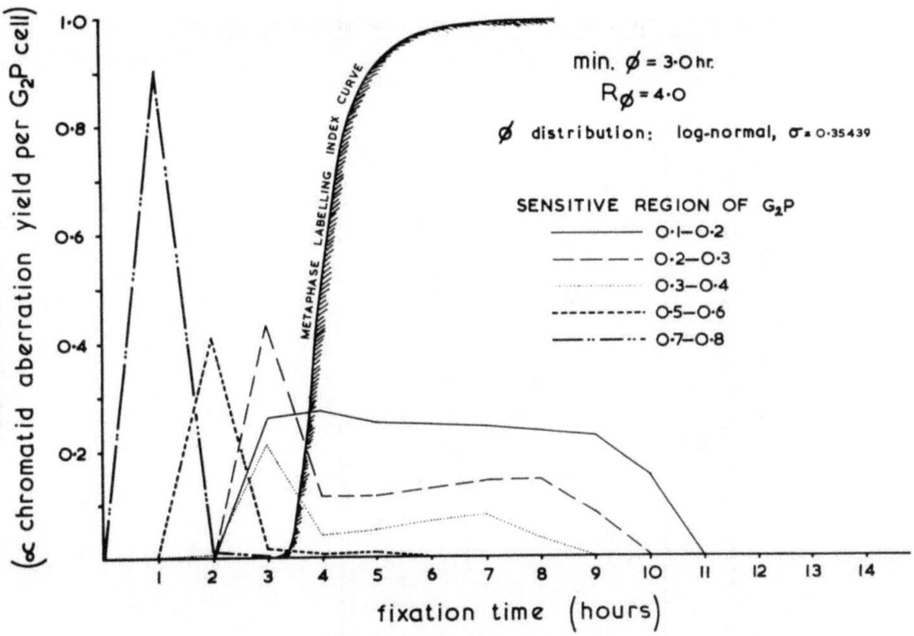

Figure 6. The effect of changing the position of the 10% sensitive region in G_2P upon the form of the aberration yield-time curve. Integral (hourly) fixation times were used.

If we keep minimum ø and Rø constant, and change the position in G_2P of the 10% sensitive region, then again there are changes in the shape of the yield-time curve, and in the position and height of the true peak frequency. This is shown in Figures 6 and 7 for the population used in Figures 2 and 4.

Figure 6 is derived from integral (hourly) fixation times, and demonstrates the increasing prominence of the "tail" as the sensitive region moves back towards the S/G_2P border.

Figure 7 gives the position and height of the true peak frequency. The order of the peaks agrees with the order of the sensitive regions in G_2P, but their positions do not correspond with those expected on the basis of the known mean G_2P. Of more consequence is the profound effect on peak aberration frequency, which varies from 0.88 aberrations per G_2P cell when the sensitive region is between 0.7-0.8 G_2P, to 0.42 per cell when the sensitive region is between 0.1-0.2 G_2P.

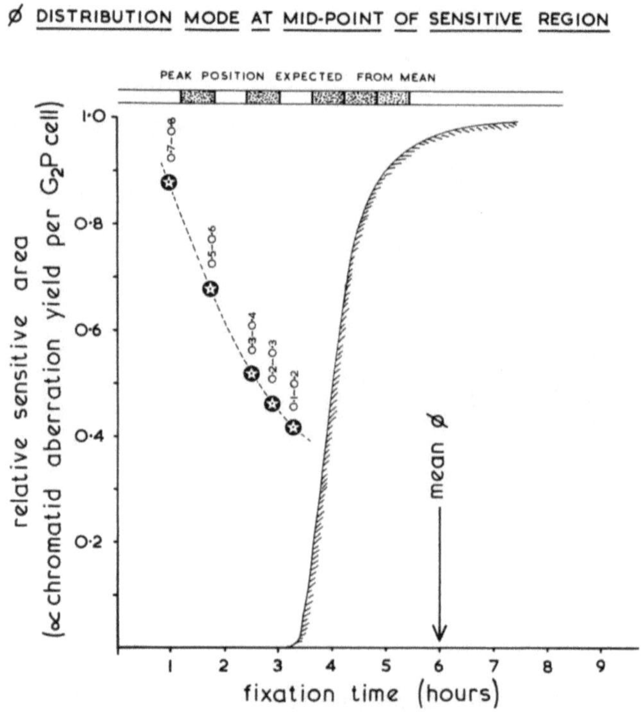

Figure 7. The effect of changing the position of the 10% sensitive region in G_2P upon the position and height of the true peak (i.e. maximum aberration frequency). The population used is the same as that for Figure 6.

We therefore conclude that in the ideal population, neither the shape of the yield-time aberration curve, nor the position of the peak in relation to the known mean Ø can be related directly to changes in radiosensitivity throughout G_2P. Moreover, since the height of the peak (= maximum observed aberration frequency) is a function of minimum Ø, RØ and the position of the sensitive region in G_2P, it cannot be used as a valid measure of radiation damage.

If we now return to the real situation in a meristem, and remember that metaphase has a finite duration, usually extended by colchicine; that radiation, internal and external, affects the values of minimum Ø and RØ by changing the shape of the Ø distribution; and that the population is undergoing considerable cell mixing during the first few hours after irradiation as evidenced by a fluctuating M.I.[8] and a variable aberration distribution pattern [6], then it is obvious that considerable care must be taken when making deductions about changes in radiosensitivity during G_2P from yield time curves of chromatid aberrations.

Moreover, any treatment given in conjunction with radiation, and which disturbs the ∅ distribution, is bound also to alter the mean frequency pattern of aberrations. Changes in yield observed after such treatments are not, therefore, necessarily indications of a direct effect upon damage or repair.

Finally, it should be emphasised that the problems outlined here for chromosomal aberrations apply with equal force to any data scored from serial post-treatment samples, and expressed on an arithmetic mean per cell basis. Labelling and mitotic index curves are examples of data likely to be affected.

REFERENCES

1. Clowes, F.A.L. (1961) Duration of the mitotic cycle in a meristem. J. exp. Bot. 12 : 283-293.
2. Evans, H.J. (1962) Chromosome aberrations induced by ionizing radiation. Int. Rev. Cytol. 13 : 221-321.
3. Lea, D.E. (1947) "Actions of radiations on living cells." (Cambridge University Press).
4. Miller, M.J. and Colaiace, J.D. (1970) Radiation dose fractionation, mitotic index, and multiple fixation studies on chromosome aberrations in *Vicia faba*. Mutation Res. 10 : 449-462.
5. Neary, G.J., and Evans, H.J. (1958) Mitotic delay induced by gamma radiation with and without oxygen in an asynchronous cell population. (Abstr.). Radiat. Res. 9 : 159.
6. Savage, J.R.K. (1970) Sites of radiation induced chromosome exchanges. Current topics in Radiation Research 6 : 129-194.
7. Savage, J.R.K., and Wigglesworth, D.J. (1970) The non-uniform radiosensitivity of the dormant root meristem of barley seed revealed by chromosome aberrations in cells at metaphase of the first division cycle. Radiat. Bot. 10 : 377-390.
8. Scott, D. and Evans, H.J. (1967) X-ray-induced chromosomal aberrations in *Vicia faba* : changes in response during the cell cycle. Mutation Res. 4 : 579-599.

ACKNOWLEDGEMENTS

The authors would like to thank Dr. R.H. Mole and Mr. D.G. Papworth for constructive criticism during the preparation of the manuscript.

This paper is partially based on work performed under contract with the U.S. Atomic Energy Commission at the University of Rochester Atomic Energy Project and has been assigned Report No. UR-49-1487.

-DISCUSSION-

DAVIDSON - Were these primary or lateral roots?

SAVAGE - They were primaries from a "cytologist's bean" where the laterals and plumules had been removed and the mitotic index is constant.

VAN'T HOF - What you have said is true, but in practical terms, the issue depends on what is given, and whilst intellectually one would like to have such precision of cell population location, I just wonder if it is necessary?

SAVAGE - That is a good point, but the trouble is that people interpret this data as if they had this precision and the purpose of the paper is to draw attention to the fact that the situation is worse than is normally assumed.

VAN'T HOF - Do you have any real objection to integrating the area under the curve?

SAVAGE - Since the values of Rϕ and ϕ are varying during sampling, the aberration curves change shape so the areas under the curve do not bear any simple relationship to the aberration yield.

HABER - Do you know of any published work on radiosensitivity of the "cell cycle" which meets all your requirements? Has it ever been done say with synchronous mammalian cells?

SAVAGE - No, I don't know of any good cytological systems where Rϕ is low enough. With regard to "synchronous" systems, it depends of course on how good the synchrony is.

VAN'T HOF - We can culture *Vicia* though as yet we have not tried to synchronize them at the G_1S boundary as in *Pisum*. If we could do this, it would be possible to irradiate the entire population all at the same stage and then release them into S. True, some

VAN'T HOF - synchrony would be lost in passing through S and G_2, but this would meet at least some of your requirements and perhaps the proportion of aberrations would be a constant.

SAVAGE - There might be difficulties when you come to interpret the curves at subsequent metaphase, but it might be worth a try.

CONGER, A. - Is there any case in which yields have been followed and the customary interpretations of changes in sensitivity applied, which you think are valid?

SAVAGE - I'm worried about this point. Firstly, if you think of the theoretical aberration curves I showed, derived from a very narrow band of sensitive cells in G_2P, many of them are strikingly similar to some published curves. Do we say, therefore, that there is a real variation in sensitivity through G_2P or that there is a very narrow sensitive region?

Secondly, in the very early fixation times after radiation, the predominant aberrations seen are a high frequency of achromatic lesions (gaps) and true structural aberrations do not appear until a later time, and at a lower frequency than gaps. Is there a real difference in sensitivity for production of gaps and structural changes or is the observed yield difference purely a function of the position in G_2P of the sensitive regions for these two types of aberrations.

LANGE - The importance of these problems depends surely upon what your assay system is. If you are looking say for reproductive integrity, many of these perturbations are irrelevant. It is only the cytologists who are going to suffer from these difficulties and almost any other endpoint will be little affected.

CONGER, A. - I'm not so sure. The problem will certainly affect labeling index interpretations like those we have had in this meeting.

SAVAGE - Also mitotic index determinations, in fact anything which is scored down a microscope and expressed as arithmetic mean events per cell. And the situation gets much worse as you go further back into the cycle from metaphase.

We are continuing theoretical work, particularly in relation to labeling index where I feel it is important. You will have noticed that in no case did the 50 percent labeling index correspond to the duration of mean ϕ, and this is not just because ϕ is a log-normal distribution. In populations with large values of $R\phi$ the 50 percent labeled index cannot be a measure of mean G_2P duration.

HABER - How reliable is the colchicine method for determining cell cycle times, i.e., the one where you give a short treatment and wait for the arrival of tetraploid cells?

VAN'T HOF - We have used this technique extensively and compared it with tritiated thymidine pulse labeling and found no difference (Expt'l. Cell Res., 41:274-288, 1966). Whatever therefore is said about the Quastler-Sherman technique must also apply to the small tetraploid cell population technique (Expt'l. Cell Res., 17:420-438, 1959).

SAVAGE - As Dr. Clowes pointed out, we can't think of the whole meristem as a single unit. I think we will have to get down to the sort of precision he uses, and look at aberration yields in the small sub-populations where values of Rø may be much smaller. We can't get much precision data from the whole root.

VAN'T HOF - When one starves *Vicia faba* root tips of carbohydrate, cells stop in G_1 and G_2 and will remain stationary for days. Yet when cell progression in the cycle is resumed cell distribution on entry into metaphase remains unchanged as compared to asynchronously dividing cells in the primary root meristem of intact seedlings.

SAVAGE - Of course the shape of the G_2P distribution curve is itself influenced by imprecision, and one wonders how much of the skewness is attributable to this problem.

LANGE - Even in a uniform population there is no relationship between the generation time of one daughter cell and that of its sister.

SAVAGE - It might be of interest to point out that the mouse gut data that Quastler and Sherman used has the lowest Rø which I have come across in the literature (Expt'l. Cell Res., 17:420-438, 1959).

BALL - I wonder if you would get similar variations in a haploid system, such as a population of *Tradescantia paludosa* microspores in an anther?

SAVAGE - Yes you do. I have data for this species and we estimate Rø to be about 4.0.

BALL - What about a system such as fern gametophyte in which all the cells are homozygous?

SAVAGE - I don't know.

VAN'T HOF - If one asks of "synchronous" bacterial systems, "How good is the synchrony?" one discovers that cell division occurs for a period of time equal to 50 percent of the total generation time. So, one wonders if we are ever going to obtain that "ideal population" of cells.

SEED MERISTEMS AS RADIOBIOLOGICAL TEST SYSTEMS*

C. F. KONZAK AND E. A. FAVRET**

Department of Agronomy & Soils and Program in Genetics
Washington State University
Pullman, Washington, 99163, U.S.A.

INTRODUCTION

Seeds or the seedling plants grown from seeds have been used as biological test material from the beginning of radiobiology[28]. Barley seeds in particular have figured in important scientific advances. Stadler (1928), for example, used barley and later maize for his experiments on the artificial mutation of the gene in plants. It is also of some significance that the same variety of barley, Himalaya (C. I. 620), treated by Stadler is even more widely used today as an international radiobiological research standard. Himalaya barley is also extensively used for physiological and biochemical studies and in chemical mutagenesis research[9, 13,15,16,27].

An advantage of seeds for mutation and other studies is their convenience as a biological material. In addition, dry "resting" seeds of many plant species can be subjected to extreme treatments with physical or chemical agents to determine the effect prior to or following exposure to a mutagen[14,15,17]. Metabolic activity in seeds can be modified over a wide range depending largely on temperature, seed moisture content and oxygen availability. Many substances are readily imbibed by seeds. When metabolism is slow, concentrations of chemicals toxic to active systems can be applied, then washed out before metabolic activity is increased, thus permitting more selective studies on many interactions. Cellular sensitivity to mutagens increases with metabolic activity and cell development. Thus, more effective and efficient mutagen treatments often can be applied to seeds than to actively growing tissues.

Seeds of the cereals, i.e., barley, wheat, oats, corn, and rye, have been widely used. The more determinate, vertical development of these monocotyledonous species offers certain advantages in a test system. Seeds of dicotyledonous species have been less used as test systems, but even with these species a seedling test provides useful data for estimating the severity of mutagen-induced damage. Thus, the small, rapidly growing crucifer, *Arabidopsis thaliana* L., has become widely used for studies of mutations induced in the seed meristem. Dry or wet weight of near full grown *Arabidopsis* plants has been used for comparing effects of mutagen treatments on growth following a seed (meristem) exposure[21,29]. The *Arabidopsis* system is particularly useful when small seed size and short generation times are prime considerations.

Experiments by numerous authors have demonstrated that the levels of injury observed in seedlings and the survival of M_1 plants following treatment with a range of doses of X rays, gamma rays, or neutrons and many chemical mutagens may be strongly correlated with the levels of other biological effects induced, e.g., mutations and chromosome aberrations[3,4,5,6,11]. The correlation has been clearest with X rays, gamma rays, and neutron irradiations, but different relationships of seedling injury to other kinds of induced damage occur for sparsely and densely ionizing radiations. The correlations between different kinds of biological effects induced in plants by certain chemical mutagens (diethyl sulfate, ethyl methane sulfonate, N-methyl and N-ethyl N'nitrosourea and sodium azide) may be lower and different from those induced by radiations[12,17,18,24]. Relationships between the levels of biological effects induced[19,24] may be appreciably altered by mutagen treatment combinations.

Since the analyses of other biological effects such as reduction in survival, chromosome aberrations, and mutations induced in seed meristems have been described in previous papers[cf 1,11], at this Conference by Conger *et al*., and in the recent IAEA "Manual of Mutation Breeding", we will not review them here.

In this paper we will describe some recent seedling development studies with the cereal (barley) seed test system. These studies have been made convenient by the now widely used "growing rack" seedling culture method of Myhill and Konzak[22]. Our new results suggest opportunities not only for further improvement of the seedling culture technique, but also for the need of further investigation of the effects of treatments on the different tissue comprising the seed meristem. This approach should lead to a better understanding of the biological mechanisms through which mutagen-induced growth inhibition is developed.

MATERIAL AND METHODS

The seed (caryopses) used in the various experiments was

barley, Hordeum vulgare L. "Himalaya" (C.I. 620), WSU strain. The seeds used for the radiation experiments were first adjusted to approximately 13-14% water content by storing them under partial vacuum over a 65% glycerol solution. The seeds treated with ethyl methane sulfonate (EMS) were taken from their dry stored condition of about 9% water content, presoaked for 18 hrs. at 2±1 C, then treated at 20±.05 C with a freshly prepared EMS solution in 0.1 M Tris buffer at pH 8.0. After treatment the seeds were washed under flowing tap water at approximately 16 C for 2 hrs., then soaked in water for 16 hrs. prior to sowing.

Gamma radiation treatments were made in the Washington State University Nuclear Radiation Center using the special underwater ^{60}Co facility developed by Botchek et al.[1]. Gamma radiation dose rates used were 4,850 to 5,100 rads/min., depending on the date of the experiment. Neutron irradiations were conducted at Washington State University's Nuclear Radiation Center by J. J. Koelling using an IAEA Standard Neutron Irradiation Facility (SNIF)[2] for which the Nuclear Reactor was operated at between 18 and 20 kilowatts power. Neutron dose estimations using TLD 600 and TLD 700 dosimeters were performed by J. J. Koelling, R. C. Brown, and W. F. Hendrickson of the Washington State University Nuclear Radiation Center. The dosimetry data obtained indicated that the barley seeds received a neutron flux of approximately 9×10^8 $n/cm^{-2}sec^{-1}$ as calculated according to procedures described by Wingate, Tochlin, and Goldstein[30]. Gamma contamination was determined to be approximately 15 R/min. of exposure in the SNIF at 20 kilowatts. Calibration of the Washington State University Standard Neutron Irradiation Facility will be the subject of a later report.

Treated and control seeds were cultured using the seedling "growing rack" technique[22] in an air conditioned laboratory maintained at 23-25 C under dim light, with approximately 70 foot candles cool white fluorescent continuous illumination. We measured an average of ten seedlings per replicate. In the neutron experiment there were 10 replicates, in the gamma experiments there were 7 replicates, and in the EMS experiment there were 5 replicates. The seedlings were measured one or more times, depending upon the experiment. In studies of daily growth, all seedlings were measured within approximately 1-1/2 hours and in the same replicate and treatment sequence beginning the fourth day after planting. The growth of each seedling and the average growth rate for seedlings in each treatment were determined. The length of the coleoptile, the first leaf blade, and the first leaf sheath were measured separately. Favret et al.[8] have previously made similar measurements to compare the growth of different mutants. In one experiment the seedlings were photographed on black and white 35 mm film to reduce the time required for recording to less than 1/2 hour and the seedlings were later measured by projecting the negative on a screen. This method proved unacceptable since neither the coleoptile nor the leaf sheath could be readily identified. We found that the plant growth in dim light made the position of the ligule difficult to distinguish. It was often necessary to dissect severely injured seedlings during their final measurement to locate the ligule and establish the length of the leaf blade and sheath. The most rapid method was to lay the pack of seedlings on a paper along a prescribed base line and to make a pencil mark on the sheet to show the point to which either the coleoptile blade or sheath had grown by the time of measurement. The seedling lengths were then measured and marked on the sheet for summarizing later. The method was time-consuming, but the time during measurement was considerably reduced. The use of ruled and numbered graph paper would have further simplified the job. More efficient methods of measurement are needed.

RESULTS

Tissue Components Measured in the Seedling

In the seedling test, it is reasonably convenient to measure the response to treatment of the coleoptile, the first leaf blade and sheath, two of the already differentiated meristems in the seed[20,23,26]. In order to clearly identify the leaf sheath and its contribution to the seedling height, it is necessary to grow the seedling until the elongation of the leaf sheath has been completed, as will be discussed. We were concerned in part whether the sheath development may differentially influence the estimate of damage measured, depending on the period of seedling development. Our results show that the coleoptile, leaf blade and leaf sheath have different response curves to neutrons (Fig. 1) as well as to gamma rays (Fig. 2) and the leaf blade and sheath appear to respond differently to treatments with the chemical mutagen ethyl methane sulfonate (Fig. 3). The "stimulation" effect observed for the leaf sheath in Fig. 1 appears to be an artifact, associated with the time of measurement. The injury response of the same set of seeds grown for 12 rather than 9 days shows no evidence of stimulation at the low dose level (Fig. 4). The 5-day seedling height measurements also result in an underestimate of the levels of damage induced by neutrons. The 5-day seedling injury response curve appears similar to the 6 and even the 9 day response curve for the leaf blade only. This suggests that the leaf blade growth was virtually complete at 5 days under our growing conditions. However, in the case of the gamma treated seeds, the room air conditioning failed between the 4th and 5th day. The higher temperature probably accelerated seedling growth sufficiently to account for the completion of blade growth at 5 days. Even so, the principles involved in the study remain the same, only the timing is slightly affected. The leaf sheath appears to be relatively more responsive than the blade to neutron irradiation, especially, but

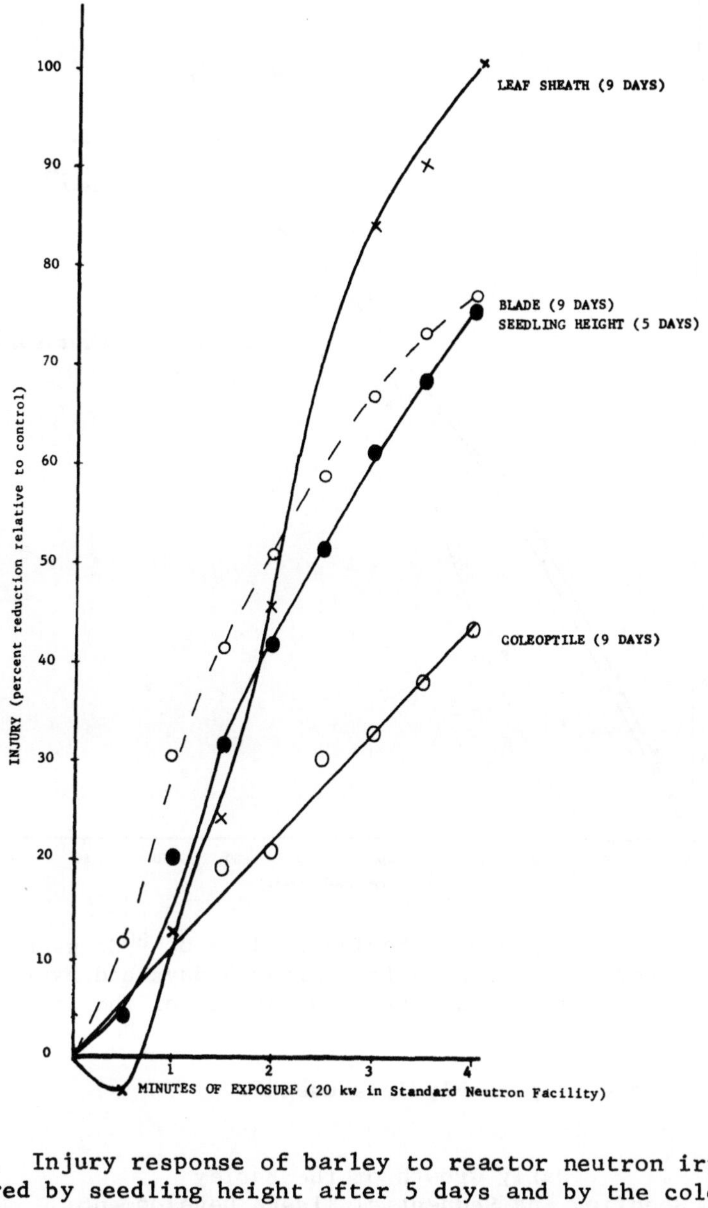

Figure 1. Injury response of barley to reactor neutron irradiation as measured by seedling height after 5 days and by the coleoptile, leaf blade and leaf sheath of the same seedlings after 9 days of growth.

is also more responsive to damage by gamma radiation and to the chemical mutagen EMS. Measurements of the leaf sheath may prove to be useful for estimating the biological effects of mutagen and other treatments. The coleoptile appears to be the least responsive of

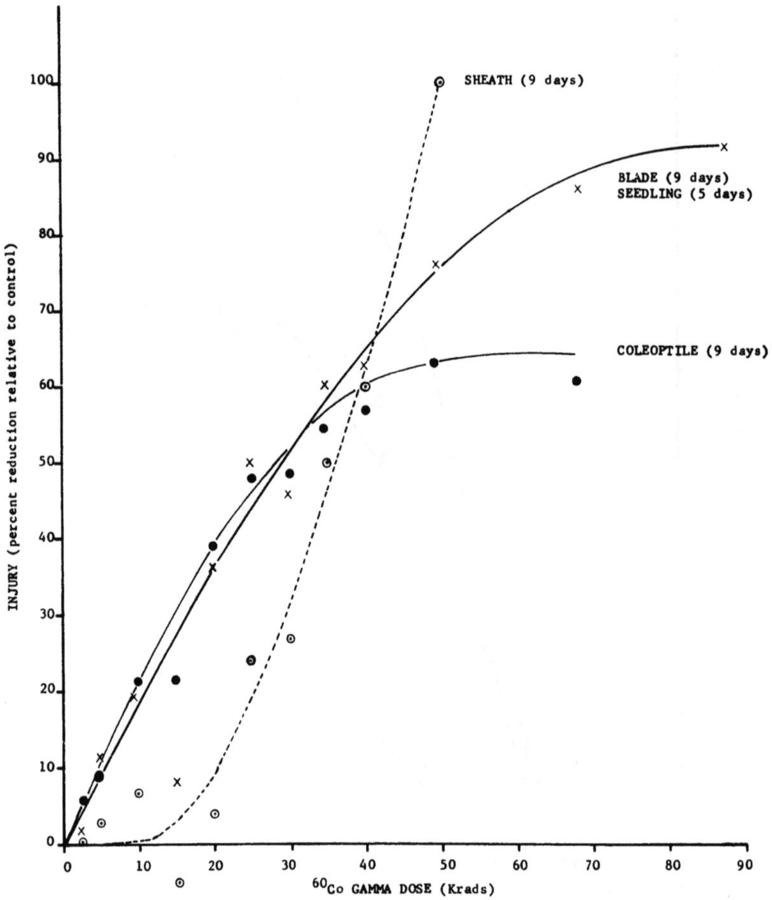

Figure 2. Injury response of barley to dose of ^{60}Co gamma irradiation as measured by seedling height after 5 days and by the coleoptile, first leaft blade and first leaf sheath of the same seedlings after 5 days of growth.

the 3 tissues and is probably less useful.

Daily Growth of the Barley Seedling and Sequential Tissue Development

Daily measurements of the development of seedlings from seeds exposed to neutrons show that the induced growth inhibition is already expressed by the leaf blade at the 4th day after planting (Fig. 5). By the 6th or 7th day the blade growth appears to be complete and sheath development has become evident especially in

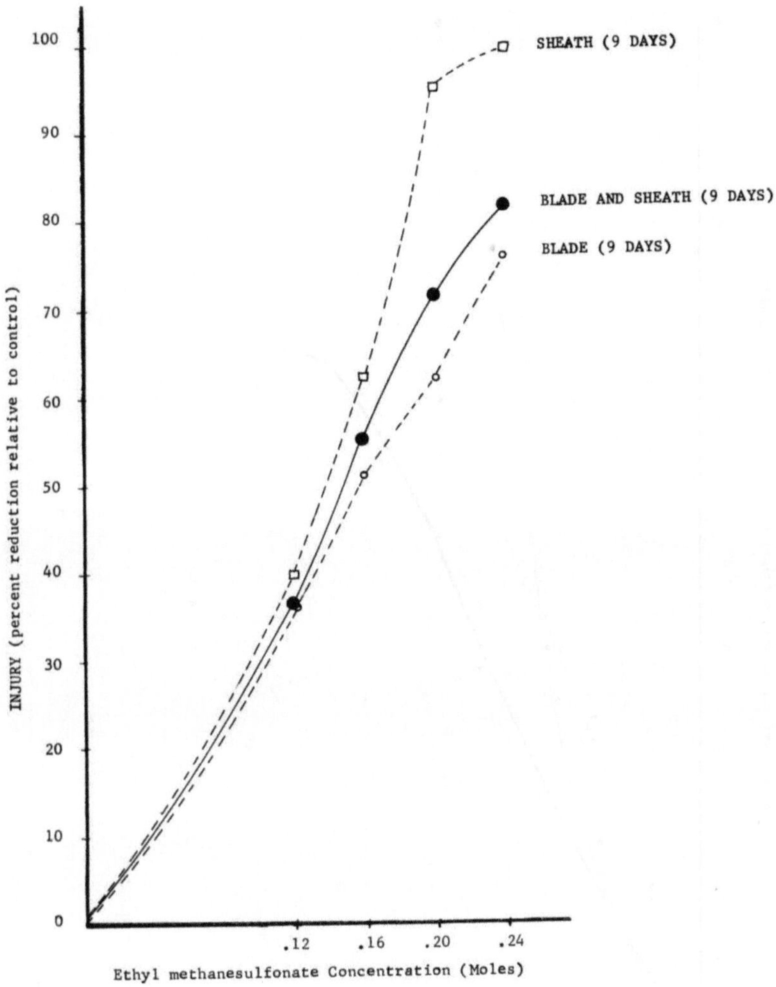

Figure 3. Injury response of the first leaf of barley and of its components at 9 days to dose of the chemical mutagen, ethyl methane sulfonate. (Seeds presoaked 16 hrs. at 2±1 C, treated 2 hrs. at 30±.05 C with EMS in 0.1 M Tris buffer at pH 8.0, post-washed 2 hrs. in flowing water at 16 C, then soaked 16 hrs. at 0 C.)

some of the seedlings from treated seeds. The data indicate that blade growth is almost completed before sheath development is able to proceed, suggesting the existence of a mechanism which controls the sequence of tissue development. This is shown by the delayed initiation of sheath growth in the control as compared with treated seedlings. However, development of the sheath was almost completely inhibited by neutron exposures longer than two minutes. The injury

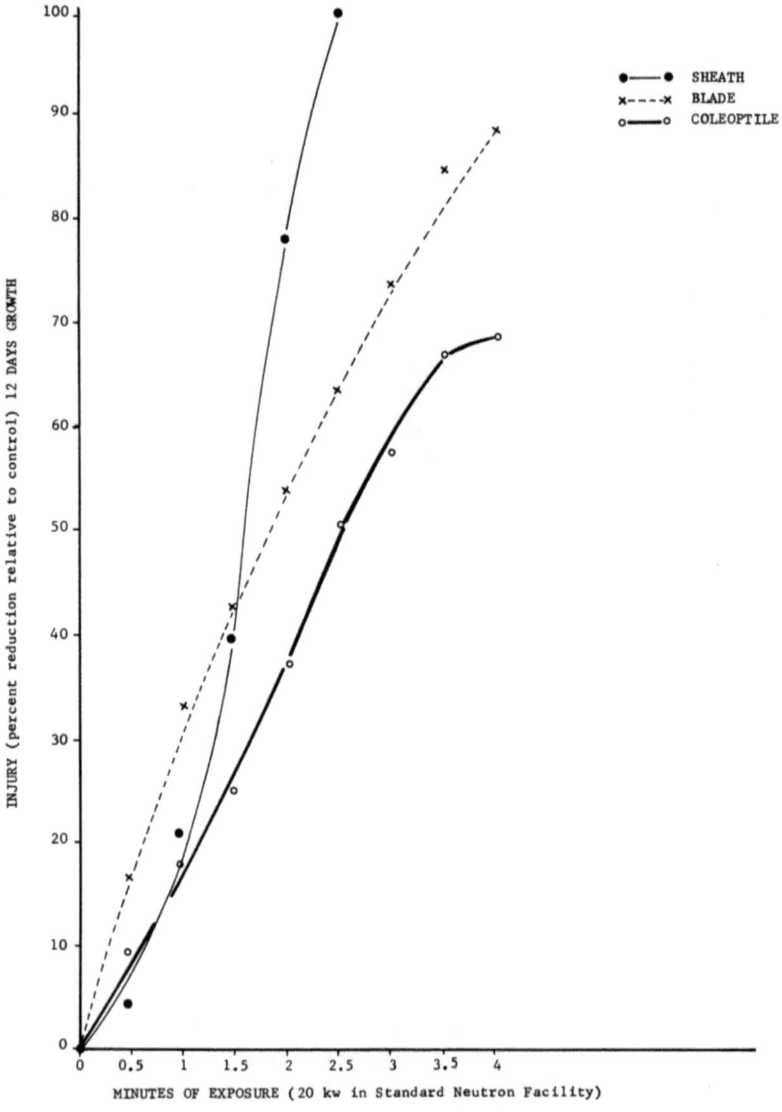

Figure 4. Injury response of barley to reactor neutron irradiation as measured by the coleoptile, first leaf blade and first leaf sheath of seedlings grown 12 days.

curves for sheath and blade intersect at about the 45% injury level. A similar relationship between leaf blade and leaf sheath development was found also for seeds exposed to gamma radiation (Fig. 6), and because of the larger number of doses applied, the series of curves show somewhat more clearly that the first leaf blade growth of

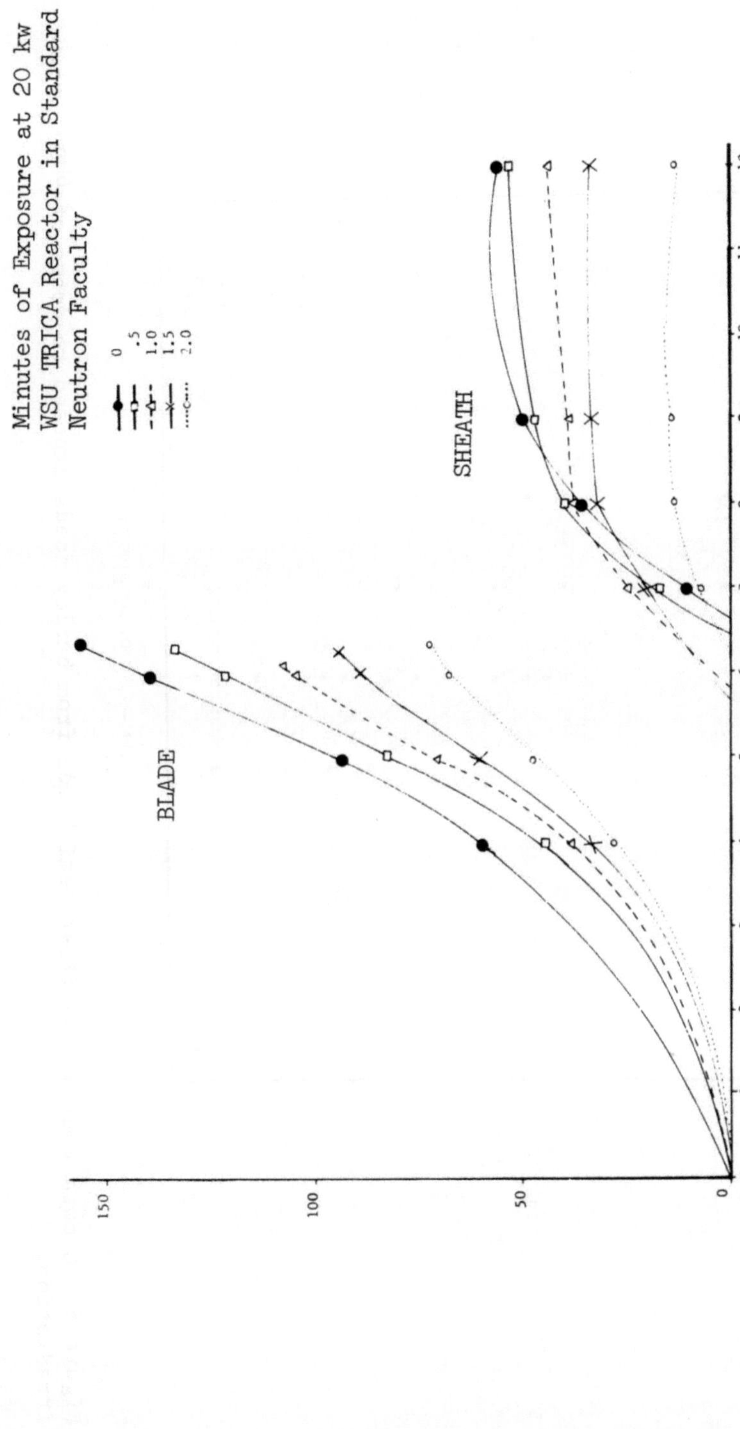

Figure 5. Growth (mm) of the first leaf blade and first leaf sheath from barley seeds exposed to several doses of reactor neutron irradiation.

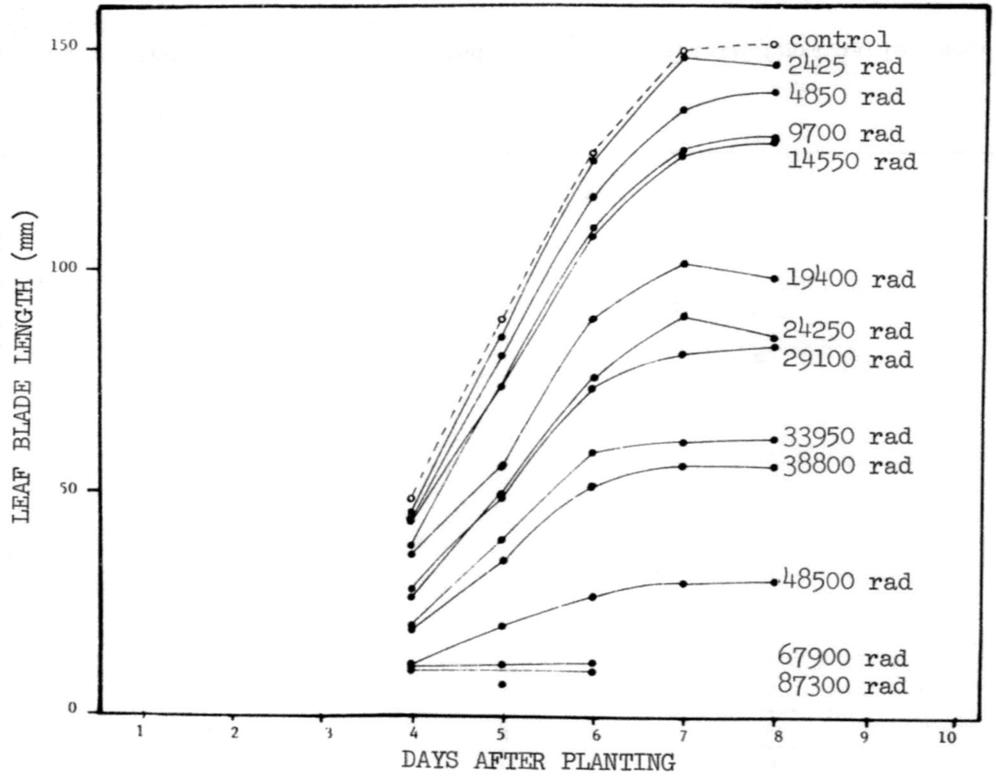

Figure 6. Growth (mm) of the first leaf blade from barley seeds exposed to doses of ^{60}Co gamma irradiation.

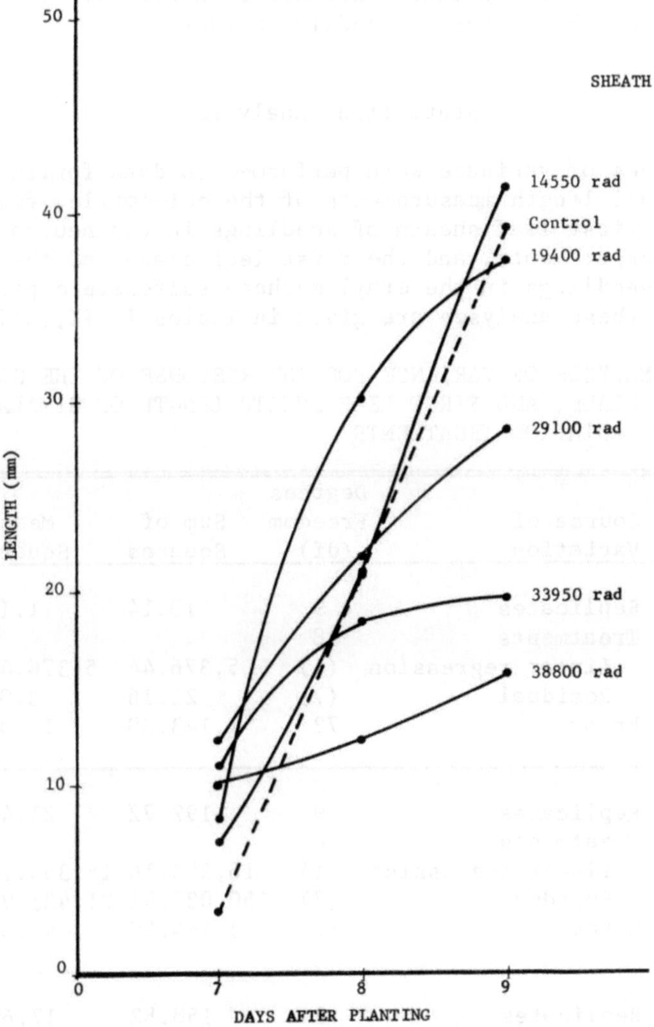

Figure 7. Growth (mm) of the first leaf sheath of seedlings from barley seeds exposed to doses of ^{60}Co gamma irradiation.

seedlings in the treatments given higher doses stops earlier than does that of the seedlings in the control and lower dose treatments. Likewise, the sheath development in the control and lower doses begins later and proceeds at a more rapid rate than is shown by the seedlings exposed to higher gamma doses (Fig. 7). It is probable that already by 6 days the sheath may be an important component of the length of the more severely injured seedlings, contributing to an underestimation of the induced damage. Since the blade growth of controls is not completed until about the 7th day, there is obviously considerable chance for error in estimating the levels of damage induced by measuring seedling height.

Statistical Analyses

Analyses of variance were performed on data for the replicate average final length measurements of the coleoptile, first leaf blade, and first leaf sheath of seedlings in the neutron and gamma radiation experiments, and the first leaf blade and the first leaf sheath of seedlings in the ethyl methane sulfonate experiments. Results of these analyses are given in Tables I, II, and III. In

TABLE I, ANALYSIS OF VARIANCE FOR THE RESPONSE OF THE COLEOPTILE, FIRST LEAF BLADE, AND FIRST LEAF SHEATH LENGTH OF HIMALAYA BARLEY TO NEUTRON RADIATION TREATMENTS

Tissue	Source of Variation	Degrees Freedom (df)	Sum of Squares	Mean Square	F Ratio
Coleoptile	Replicates	9	10.14	1.13	<1.0
	Treatments	8			
	Linear regression	(1)	5,376.44	5,376.44	338.59**
	Residual	(7)	21.16	3.02	<1.0
	Error	72	1,143.30	15.88	
Blade	Replicates	9	192.72	21.41	<1
	Treatments	8			
	Linear regression	(1)	16,353.16	16,353.16	374.47**
	Residual	(7)	150,037.91	21,433.98	490.82**[1]
	Error	72	3,144.50	43.67	
Sheath	Replicates	9	158.82	17.65	<1
	Treatments	4			
	Linear Regression	(1)	1,119.36	1,119.36	44.83**
	Residual	(3)	11,092.71	3,697.57	148.08**[1]
	Error	33	824.21	24.97	

[1]This statistically significant variation indicates the existence of second and higher order regressions.

Table IV, the linear regressions were calculated by the least squares method. The analyses show that differences attributable to treatment and to linear regression on dose are all highly significant statistically. The variation due to linear regression on dose

TABLE II, ANALYSIS OF VARIANCE FOR THE RESPONSE OF THE COLEOPTILE, FIRST LEAF BLADE, AND FIRST LEAF SHEATH LENGTH OF HIMALAYA BARLEY TO GAMMA RADIATION TREATMENTS

Tissue	Source of Variation	Degrees Freedom (df)	Sum of Squares	Mean Square	F Ratio
Coleoptile	Replicates	6	23.82	3.97	1.364
	Treatments	11	2,830.85		
	Linear regression	(1)	336.20	336.20	115.533**
	Residual	(10)	2,494.65	249.46	85.725**
	Error	66	191.81	2.91	
Blade	Replicates	6	156.44	26.07	<1.0
	Treatments	9	114,756.88		
	Linear regression	(1)	15,894.72	15,894.72	83.341**
	Residual	(8)	98,862.16	12,357.77	64.795**
	Error	54	10,298.94	190.72	
Sheath	Replicates	6	531.83	88.63	3.74
	Treatments	9	7,925.39		
	Linear regression	(1)	4,843.95	4,843.95	204.386**
	Residual	(8)	3,081.44	385.18	16.252**
	Error	54	1,279.81	23.70	

TABLE III, ANALYSIS OF VARIANCE FOR THE RESPONSE OF THE FIRST LEAF BLADE AND FIRST LEAF SHEATH LENGTH OF HIMALAYA BARLEY TO EMS TREATMENTS

Tissue	Source of Variation	Degrees Freedom (df)	Sum of Squares	Mean Square	F Ratio
Blade	Replicates	9	509.86	56.65	<1.0
	Treatments	4			
	Linear regression	(1)	11,784.08	11,784.08	141.25**
	Residual	(3)	106,825.33	35,608.44	426.81**
	Error	36	3,003.50	83.43	
Sheath	Replicates	9	72.06	8.01	<1.0
	Treatments	2	5,209.30		
	Linear regression	(1)	508.46	508.46	28.33**
	Residual	(1)	4,700.84	4,700.84	261.89**
	Error	18	323.18	17.95	

TABLE IV, LINEAR REGRESSION COEFFICIENTS (b) FOR TISSUE LENGTH/DOSE FOR NEUTRON, GAMMA RAY, AND EMS TREATMENTS OF HIMALAYA BARLEY SEEDS

Treatment	Tissue	Absolute Value	Relative Value (% mean)
Neutrons	Coleoptile	b = - 6.0 mm min^{-1}	-29.2%
	Blade	b = -33.0 mm min^{-1}	-44.2%
	Sheath	b = -21.2 mm min^{-1}	-54.1%
Gamma rays	Coleoptile	b = -0.262 mm Kr^{-1}	- 1.6%
	Blade	b = -2.992 mm Kr^{-1}	- 2.2%
	Sheath	b = -0.974 mm Kr^{-1}	- 2.8%
EMS	Blade	b = -58.96 mm 10^{-4} M 2 hr., 20 C	-71.0%
	Sheath	b = -19.13 mm 10^{-4} M 2 hr., 20 C	-58.5%

accounts fully for the differences due to treatment only for the coleoptile in the neutron treatments. In all other cases there was in addition a statistically highly significant residual non-linear component of variation, indicating the presence of second and higher order interactions. No effort was made to extend the analysis to further subdivide the residual component of variation due to treatments. The variation due to replicates, however, was less than the experimental error.

From the linear regression coefficients both absolute and relative reductions in coleoptile, leaf blade, and leaf sheath length per unit of dose were calculated from the data of the neutron and gamma radiation experiments and from the blade and sheath length data of the EMS experiment (Table IV). The relative values for biological effects induced per unit dose of radiation illustrate rather clearly the relatively greater responsiveness of the sheath and the relatively lower responsiveness of the coleoptile to radiation. In contrast, the leaf blade appears somewhat more responsive per unit dose to EMS than does the leaf sheath. Interestingly enough, however, when the dose/effectiveness ratios for the coleoptile and the first leaf sheath in the neutron and gamma experiments are compared, the values obtained are almost identical and about double that obtained for the blade, suggesting that the blade is relatively less responsive to neutrons than to gamma rays (Table V). In contrast, the effectiveness ratios for the EMS and gamma radiation treatments are identical, indicating that the blade and sheath respond similarly to EMS and gamma radiation.

Statistical analyses of data from the daily growth measurements of the seedlings will be the subject of a later paper. The results

TABLE V, RATIOS OF DOSE/EFFECTS INDUCED BY NEUTRONS, GAMMA RADIATION AND EMS ON THE COLEOPTILE, FIRST LEAF BLADE, AND FIRST LEAF SHEATH OF HIMALAYA BARLEY

Tissue	Ratio between effects per unit dose		
	Neutrons (1 min.)/γ-rays (1 Kr)		
Coleoptile	6.0/0.262	=	22.90
Blade	33.0/2.992	=	11.03
Sheath	21.2/0.974	=	21.77
	EMS (10^{-3} M 2 hr. 20 C)/γ-rays (1 Kr)		
Blade	58.96/2.992	=	19.71
Sheath	19.13/.974	=	19.64

from these measurements suggest, however, that daily growth data may provide another useful basis for comparing different treatments, particularly when it is difficult to obtain a suitable growth end point.

DISCUSSION AND CONCLUSIONS

It should be recalled that the meristem of the barley seed is a complex tissue composed of from 3 to 5 tiller primordia in various stages of development. The coleoptile is fully differentiated in the mature embryo. Most of its growth is due to elongation, relatively little is due to increase in cell number. However, the meristem of the first leaf grows by both cell division and elongation. However, there are two reasons to suppose that this meristem is already differentiated into its two components, the blade and the sheath. First is their sequential development and second is their differential response to mutagen and other treatments. Because of the sequential development and the differential treatment responsiveness of these two seedling height components, it is important that at a minimum, measurements are taken after their growth has terminated. Thus, if the first leaf blade is measured, the seedling should be grown at least 6 to 7 days and any growth of the sheath should be discounted. If the first leaf sheath, the total first leaf length, or both the sheath and the blade are to measured, the seedlings must be grown for 12 or more days. Otherwise, the estimate of induced biological effects will be biased. We found that mutagen-induced injury was underestimated when seedling height measurements were made between 5 and 12 days, but it is possible that injury expressed by the leaf sheath alone may correlate better with some other biological effects (for example, reduction in survival than either injury expressed by the leaf blade or the total seedling height). The second leaf should not be included in the seedling height measurement since this leaf is another meristem and may have different response characteristics. Our studies indicate that seedling height by itself is not a suitable biological measure to compare rates or frequencies of different biological effects induced in the first leaf meristem.

Growth rate analyses provide information on the time of growth initiation in the different tissues of the leaf, the growth rate of these tissues and the time at which growth in each tissue terminates. Differential responses may be expressed in any of these three aspects of growth dynamics.

Since the dynamics of growth in different tissues may be differentially affected by treatments, the analysis of growth rates could be particularly useful in investigations, comparing effects of environments, genotypes, chemicals or radiations on meristems. Growth rate analyses may reveal developmental differences which otherwise could lead to false conclusions, as for example, the apparent stimulatory effect on sheath elongation observed in our experiments was explained by the different time of initiation of growth in the sheath provoked by the treatments.

SUMMARY

1. In the dormant barley seed the meristem which forms the first leaf is already differentiated into the blade and the sheath.

2. The growth of the first leaf blade and first leaf sheath is sequential in that order. On the contrary, the coleoptile growth appears to be independent.

3. Under our experimental conditions, growth of the first leaf blade was completed by the 6th or 7th day from planting, while the growth of the first leaf sheath was not completed until the 12th day.

4. Mutagen treatment differentially alters the time of growth initiation and termination as well as the growth rate of the blade and the sheath of the first leaf.

5. In the dormant seed, the meristem originating the coleoptile is less sensitive to mutagen treatments than the meristem initial of the first leaf.

6. The blade and the sheath of the first leaf have different responses to mutagen treatments. The blade appears more sensitive in the lower dose range while the sheath is more sensitive in the higher dose range. At about 50% injury the first leaf

blade and sheath appear to respond differentially to neutrons as compared with gamma radiation and EMS. However, the coleoptile and first leaf sheath have similar reactions to the two radiations.

7. The different parameters measureable on the seedling, i.e. the final length of the first leaf blade and the first leaf sheath, and the growth rate of each tissue may prove useful in distinguishing effects of different treatments applied to seeds.

REFERENCES

1. Botchek, C. M., Konzak, C. F., and Nilan, R. A. (1963) Design of the Co^{60} facility at Washington State University. Report at the 1963 Electro-Nuclear Conference, Institute of Electrical & Electronic Engineers, Richland, Washington. C.P.A. 63-5139, Engineering Paper No. 1604. Douglas Missile and Space Systems Division. 19 p.

2. Burtscher, A. (1968) Experience with the standard neutron irradiation facility in the Astra reactor. pp. 97-106. In: Neutron Irradiation of Seeds II, Technical Reports Series No. 92 (Second Report of an FAO/IAEA Meeting on Co-ordination of research on the Use of Neutrons in Seed Irradiation, Vienna, Dec. 11-15, 1967). IAEA, Vienna.

3. Caldecott, R. S. (1955) The effects of X-rays, 2-MEV electrons, thermal neutrons, and fast neutrons on dormant seeds of barley. Annals of the New York Academy of Sciences 59:514-535.

4. Caldecott, R. S., Beard, B. H., and Gardner, C. O. (1954) Cytogenetic effects of X-ray and thermal neutron irradiation on seeds of barley. Genetics 39:240-259.

5. Caldecott, R. S., Frolik, E. G., and Morris, Rosalind (1952) A comparison on the effects of X-rays and thermal neutrons on dormant seed of barley. Proc. of Nat. Acad. Sci. 38:804-809.

6. Ehrenberg, L., Gustafsson, Å., and Nybom N. (1952) Effects of ionizing radiations in barley. Arkiv för Botanik 1:557-568.

7. Ehrenberg, L. and Nybom, N. (1954) Ion density and biological effectiveness of radiations. Acta Agriculturae Scandinavica 4:396-418.

8. Favret, E. A., Ryan, G. S., and Malvarez, E. M. (1969) Mutaciones inducidas que afectan al crecimiento inicial de la cebada. pp. 109-121. In: Induced Mutations in Plants (Proc. of the

FAO/IAEA Symposium on the Nature, Induction and Utilization of Mutations in Plants, Pullman, Washington, 1969). IAEA, Vienna.

9. Filner, P. and Varner, J. (1967) A test of de novo synthesis of enzymes: Density labeling with H_2O^{18} of barley α-amylase induced by gibberellic acid. Proc. Nat. Acad. Sci. 58:1520-1526.

10. Froese-Gertzen, Edith E., Konzak, C. F., Nilan, R. A., and Heiner, R. E. (1964) The effect of ethyl methanesulfonate on the growth response, chromosome structure and mutation rate in barley. Radiation Botany 4:61-69.

11. Gaul, Horst (1964) Mutations in plant breeding. Radiation Botany 4:155-232.

12. Heiner, R. E., Konzak, C. F., Nilan, R. A., and Legault, R. R. (1960) Diverse ratios of mutations to chromosome aberrations in barley treated with diethyl sulfate and gamma rays. Proc. Nat. Acad. Sci. 46:1215-1221.

13. International Atomic Energy Agency (1970) Mutagen effects observable in the first generation. pp. 85-106. In: Manual on Mutation Breeding, Technical Reports Series No. 119. IAEA, Vienna.

14. Konzak, Calvin F. (1957) III. Genetic effects of radiation on higher plants. The Quarterly Review of Biology 32:27-45.

15. Konzak, C. F., Bottino, P. J., and Nilan, R. A. (1968) Irradiation of seeds: A review of procedures employed at Washington State University. pp. 83-96. In: Neutron Irradiation of Seeds II, Technical Reports Series No. 92. IAEA, Vienna.

16. Konzak, C. F., Mikaelsen, K., and Sigurbjörnsson, B. (1967) Recommended standard procedures for irradiating, cultivating and measuring cereal seeds to determine the effects of neutron irradiation in the neutron-seed-irradiation programme. pp. 103-107. In: Neutron Irradiation of Seeds, Technical Reports Series No. 67. IAEA, Vienna.

17. Konzak, C. F., Nilan, R. A., Harle, J. R., and Heiner, R. E. (1961) Control of factors affecting the response of plants to mutagens. pp. 128-157. In: Fundamental Aspects of Radiosensitivity, No. 14. Brookhaven Symposia in Biology, Brookhaven National Laboratory, Upton, New York.

18. Konzak, C. F., Nilan, R. A., Wagner, J., and Foster, R. J. (1965) Efficient chemical mutagenesis. pp. 49-70. In: The Use of

Induced Mutations in Plant Breeding, Suppl. to Radiation Botany, Vol. 5. Pergamon Press Ltd.

19. Konzak, C. F., Wickham, Irene M., and M. J. de Kock (1971) Advances in methods of mutagen treatment. Paper prepared for Latin American Study Group Meeting on Induced Mutations and Crop Improvement at Buenos Aires and Castelar, Argentina, November 16-20, 1970. IAEA, Vienna. In Press.

20. Lunden, A. O. (1964) Seed embryo features and irradiation response. Radiation Botany 4:429-437.

21. McKelvie, A. D. (1963) Studies in the induction of mutations in Arabidopsis thaliana (L.) Heynh. Radiation Botany 3:105-123.

22. Myhill, R. R. and Konzak, C. F. (1967) A new technique for culturing and measuring barley seedlings. Crop Science 7:275-276.

23. Mullenax, R. H. and Osborne, T. S. (1967) Normal and gamma-rayed resting plumule of barley. Radiation Botany 7:273-282.

24. Nilan, R. A. (1971) Mutagenic specificity in flowering plants: Facts and prospects. Paper prepared for Latin American Study Group Meeting on Induced Mutations and Crop Improvement at Buenos Aires and Castelar, Argentina, November 16-20, 1970. IAEA, Vienna. In Press.

25. Nilan, R. A., Konzak, C. F., Wagner, J., and Legault, R. R. (1965) Effectiveness and efficiency of radiations for inducing genetic and cytogenetic changes. pp. 71-89. In: The Use of Induced Mutations in Plant Breeding, Suppl. to Radiation Botany, Vol. 5. Pergamon Press Ltd.

26. Osborne, T. S. and Lunden, A. O. (1965) Prediction of seed radiosensitivity from embryo structure. pp. 133-149. In: The Use of Induced Mutations in Plant Breeding, Suppl. to Radiation Botany, Vol. 5. Pergamon Press Ltd.

27. Sparrow, A. H., Bottino, P. J., and Schairer, L. A. (1970) The use of higher-plant test systems for chemical mutagenesis. Brookhaven National Laboratory Report. BNL 14011.

28. Sparrow, Arnold H., Binnington, John P., and Pond, Virginia (1958) Bibliography on the effects of ionizing radiations on plants, 1896-1955. 222 pp. Biology Department, Brookhaven National Laboratory, Upton, New York.

29. Smith, H. H. and Hirono, Y. (1969) Responses in mutation, growth inhibition, tumorization and isozyme multiplicity from

exposing seeds to irradiations of different linear energy transfer. pp. 231-243. In: Induced Mutations in Plants. IAEA, Vienna.

30. Wingate, C. L., Tochlin, E., and Goldstein, N. (1967) Response of lithium fluoride to neutrons and charged particles. pp. 421-434. In: Luminescence Dosimetry, ed. by Frank H. Attix. U. S. Atomic Energy Commission.

*Scientific Paper No. 3755. College of Agriculture and Research Center, Washington State University, Pullman, Projects 1068 and 1746 and Publication No. GEN 461 of the Instituto de Fitotecnia, C.N.I.A.I.N.T.A., Castelar, Argentina. Supported in part by the U. S. Atomic Energy Commission Contract AT(45-1)-2221. AEC Paper RLO-2221-T2-9.

**Professor of Agronomy and Genetics, Department of Agronomy and Soils and Program in Genetics; Director, Institute of Plant Breeding, Instituto Nacional de Tecnologia Agropecuaria, Castelar, Argentina and National Science Foundation Senior Foreign Scientist Fellow, Program in Genetics, Washington State University, Pullman, Washington, 99163, U.S.A., respectively.

-DISCUSSION-

CONGER, B. - When you grow your seedlings for 12-14 days do you supply nutrient solutions? I was wondering if the food supply is running out at this time and the plateau you observe is a consequence of this?

KONZAK - They were in distilled water, but we are quite sure that the food supply has not run out. One can, however, modify the response by adding nutrients to the water. We have done experiments with gibberelic acid solutions, and a differential response is obtained. The sheath is affected more than the leaf; the coleoptile growth is unaffected.

CONSTANTIN - What is the RBE for neutrons as measured by the coleoptile because this is essentially a non-dividing system? Does it approach the values Dr. Haber reported for his gamma plantlets?

KONZAK - I don't know; perhaps Dr. Haber could answer this?

HABER - The coleoptile is a dividing system at the time Dr. Konzak irradiates. It only becomes non-dividing at a later stage. Rose and Adamson have studied cell numbers and it increases about three or four-fold, so there is quite a bit of division involved (Planta, 88:274-281, 1969).

CONSTANTIN - I notice you now recommend growing seedlings for 12 days rather than five days which is the time used in much of your previous work. What does this imply?

KONZAK - It implies that in general we have tended to underestimate the amount of damage that radiation induces. The underestimate is slight, and results primarily from the fact that the control is still growing at five days. The control blade stops growing at around six days. All parts of the severely irradiated seedlings have stopped growing before this, but at the lower doses, growth continues for longer and longer periods.

CONGER, B. - What is the height of the control seedlings after 12 days?

KONZAK - Twenty to 24 cm. for the total height.

CONGER, A. - It sounds as if your sheath meristem is behaving like a "quiescent center," remaining dormant until released into activity by cessation of blade growth. Is this so?

KONZAK - Our data indicate that growth is sequential, and when the blade stops growth, the sheath can begin to elongate. However, we have not done any labeling experiments to investigate synthesis patterns in these tissues.

DAVIDSON - I would like to make a comment on coleoptiles. I think it would be dangerous to generalize from the work of Rose and Adamson which was done on selected material (Planta, 88:274-281, 1969). There are coleoptiles which occur in Barley which show no growth and almost no increase in cell number over a four-day period. It is convenient to ignore them, but nevertheless they are still part of the population. Samples should be selected at the beginning and not at the end.

KONZAK - Yes, there is variation. And we have found that different genotypes may respond differently, particularly with regard to the radiation response of the coleoptile.

D'AMATO - I think that histogenetic markers, e.g., heterozygosity for chlorophyll mutations, would be very useful in this material where you have different meristems in leaf and sheath. A study of induction, frequency, and distribution of chlorotic spots would be most interesting. Have you done anything like this?

KONZAK - No, but Dr. Smith has made such measurements with maize, though at a later stage of development. I am just not sure that one would see anything in this first leaf sheath, the patches would be very small, in some cases just a few cells. I think that the leaf

KONZAK - blade would be better, since any altered tissue can be scored microscopically in terms of individual cells or patches affected.

OSBORNE - How many different meristems do you feel that you are testing with this system?

KONZAK - Only two if the coleoptile can be considered a true meristem. This doesn't mean one couldn't measure the growth of additional leaves, but in this system the seedlings are pretty weak after 12 days, and one would probably have to go to soil culture.

I would like to emphasize again the value of repeated measurements during growth. There is a lot more information available from such data.

CONSTANTIN - We have done repeated measurement studies of total plant height of irradiated soy bean seedlings. The control plants have the typical sigmoidal growth curve. Every increment of radiation reduced total height, but the sigmoidal growth curve was retained, the plateau being reached earlier. So the whole organism here is showing a similar response to that which you are showing for a single leaf.

KONZAK - This emphasizes the error one could have from single growth measurements when the plants are still growing. The trouble is that repetitive measurements of many seedlings are very time consuming. We have tried many techniques, including static photography, and have considered time-lapse cinematography. The difficulty lies in the resolution one can achieve; it just is not possible to see all the parts one wants to measure. For example, the ligule is somewhat difficult to identify in some materials. Light especially affects the appearance of anthocyanins in the sheath, and the coleoptile tip is on the opposite side of the leaf sheath from the ligule opening.

SAVAGE - Could I add a caution about time-lapse photography. Mr. Wigglesworth in my laboratory has found in Campelia that measurements of daily flower opening patterns from time-lapse films are quite different from manually obtained data, and suspects that although intermittent light is used the intensity of it required to get the photographic images is sufficient to affect photoperiodic response.

HABER - Can I echo the caution. We studied wheat coleoptile elongation by putting Indian ink spots on different parts and showed that these parts could respond differentially to different light treatments such that no overall elongation in growth occurred (Plant Physiol., 38:523-539, 1963).

KONZAK - We have also observed differential growth including that of individual cells in different parts of barley coleoptiles. This was done by a Vietnamese trainee, Mr. N. VanMung, in our laboratory last year. He found that the cells of the upper third of the coleoptile epidermis layer elongated more than did those in the lower or middle third.

CONTRIBUTIONS OF SEED MERISTEMS TO RADIOBIOLOGY*

B. V. Conger

University of Tennessee, Department of Agronomy

Knoxville, Tenn., and UT-AEC Agricultural

Research Laboratory †, Oak Ridge, Tenn.

INTRODUCTION

"Dry" or dormant biological systems such as bacterial spores, cysts of brine shrimp and seeds of higher plants have been extensively used as test organisms in radiobiological studies. In particular, experiments conducted with the bacterial spore have contributed much to our knowledge of certain radiobiological concepts (see ref. 39 for a review). Plant seeds have been used in similar studies and although inferior to the bacterial spore in population size which can be studied, the seed has several advantages. It represents a multicellular organism with a complex chromosome structure and in addition to reduced survival (the only kind of damage studied in the spore) other radiobiological endpoints such as mutations, chromosome aberrations and growth reduction can be readily measured.

The seed is a very unique stage in the life cycle of higher plants because cell division and respiration are almost nil. Hence, it tolerates extreme physical conditions. Radiobiological experiments with seeds have been conducted for two main purposes.

---

*Published with the permission of the Dean of the University of Tennessee Agricultural Experiment Station, Knoxville, Tennessee.

† Operated by the Tennessee Agricultural Experiment Station for the U. S. Atomic Energy Commission under Contract No. AT-40-GEN-242. Research supported in part by Hatch Project TEN00287.

First, for the purpose of mutation breeding, experiments have been conducted to determine which dose or range of doses to apply to a particular species[30,36] and which modifying treatments will result in the most effective and efficient rates of mutation induction[35]. Second, experiments have been conducted to advance our knowledge on certain mechanisms (especially the early events involving reactions of free radicals) by which ionizing radiations induce damage in living cells.

This paper will be concerned primarily with the latter and specifically the effects of X- and γ-radiation. It is from these studies that seeds have made the biggest contribution to the present understanding of radiobiology. Studies on the effects of neutrons on seeds have not yet contributed as much basic information. Due to the short length of this paper and the many papers which have been published dealing with the effects of radiation on seeds a complete literature review is not possible. Rather, important works will be mentioned in the text where appropriate. Most of the information gained has come from experiments dealing with the role of modifying factors. The data presented will deal with the interrelationship between these factors and the participation of radiation-induced free radicals in the enhancement of radiation damage, especially that which is oxygen-dependent.

REACTIONS OF RADIATION-INDUCED FREE RADICALS IN BIOLOGICAL SYSTEMS

According to Powers[38], the postulate that part of the damage induced by high energy radiations in biological systems results from the formation of free radicals has been entertained for many years. Proof of this, however, has been presented for only a few biological systems. In "wet" or actively metabolizing tissues it is unlikely that free radicals persist for more than 10^{-5} sec.[2] They react either by combination with another radical or with another molecule possessing unpaired electrons.

Oxygen, a biradical, can readily react with organic free radicals to form peroxy radicals. Increased biological damage results either by the irreversibility of the change in molecules which have become peroxy radicals or by the oxidation of other (normal) molecules by the peroxy radicals. Bacq and Alexander[2] have summarized these phenomena as follows:

$$\text{RH} \xrightarrow{\text{exposed to radiation}} \text{R}\cdot + \text{H}\cdot \xrightarrow{\text{recombination}} \text{RH}$$

$$\text{Radical combination } \text{R}\cdot + \text{R}\cdot \longrightarrow \text{R-R}$$

These reactions lead to a reduced amount of biological damage.

However, if the free radical combines with oxygen

$$R^{\bullet} + O_2 \longrightarrow RO_2^{\bullet} \text{ peroxy radicals}$$

increased biological damage results.

In addition to combination or recombination of radicals, peroxy radical formation may also be prevented by exposure to protective chemical agents or radical scavengers before exposure to oxygen. Nitric oxide possesses one unpaired electron and can interact readily with organic radicals to form RON.[9,21] Hydrogen sulfide protects by donating an H atom to the free radical to form RH.

To study reactions involving radiation-induced free radicals in vivo, a biological system which can be subjected to very low temperatures and/or extreme desiccation is needed so that the free radicals can be trapped. Seed embryos from many plant species will tolerate these extreme treatments and the conditions of seed water content and temperature can be varied such that the decay of free radicals can be studied over a specified period of time post-irradiation.

Radiation-induced free radicals in seeds were first discovered by Zimmer et al.[41] using electron paramagnetic resonance (EPR) techniques and later studies confirmed this observation.[11,22] Many workers feel that free radicals induced in seeds by ionizing radiations arise from large organic molecules. Modifying factors, such as temperature and changes in seed water content, regulate radical mobility.[9,21] When the material is very dry radicals may be trapped and retained almost indefinitely. The addition of water provides a medium in which radicals can move and combine with each other. Similarly, increasing temperature during or after irradiation enhances radical mobility and hence decay. Increased radical decay during anoxic irradiation results in reduced post-irradiation oxygen-dependent damage.

MATERIALS AND METHODS IN SEED RADIO-BIOLOGICAL STUDIES

Most experiments dealing with the effects of ionizing radiations on seeds and in those to be described here have utilized the caryopsis (hereafter referred to as seed) of barley Hordeum vulgare L. emend Lam. and particularly the c.v. Himalaya as the test organism. This cultivar also serves as the biological monitor for the FAO/IAEA coordinated Program of Research on the Use of Neutrons in Seed Irradiation. Techniques for adjusting and measuring seed water content, irradiation and post-irradiation atmosphere and temperature control have been published.[18,26,34] Basically, seeds are irradiated in vacuo in sealed glass vials. After irradiation they are soaked either immediately in oxygen- or nitrogen-bubbled water at 0° C for 16-18 hours or stored for a desired period before soaking. By utilizing the technique of irradiation *in vacuo*,[14] seed water content and atmospheric conditions can be precisely controlled during and after irradiation even for post-irradiation storage periods ranging up to several weeks.[16] This technique, coupled with controlled atmospheric conditions during post-irradiation soaking, allows for the clear separation of oxygen-dependent from oxygen-independent events.

Biological endpoints most commonly measured, include reduction in seedling growth, chromosome abnormalities (both mitotic anaphase bridges and fragments and interchanges at meiosis) and frequencies of mutations scored in the M_2 generation. A good correlation exists between reduction in seedling growth and cytogenetic and genetic damage.[4,10,18,34] Since reduction in seedling growth is the easiest and most rapid endpoint to obtain, it has been used most frequently. In barley, this consists of measuring the height of the first leaf after 5-7 days growth in blotter cultures.[31] The first leaf is already formed in the embryo and growth after seed germination is the result of both cell elongation and division.

RESULTS AND DISCUSSION

As stated earlier, most of the experiments with seeds which have advanced our knowledge in radiobiology have dealt with the effects of modifying factors. The most important include atmosphere, seed water content, temperature and post-irradiation storage. Prior to 1960, most experiments were designed to study the role of these factors independently and little was known about their interrelationship in the mediation of radiation damage. Thus, the general opinion was that these factors were more or less equally important in the modification of radiation damage. However, it has been found that oxygen is a major factor and the effects of other factors secondary, their main role being that of modifying the oxygen effect. The following discussion will review and relate results of the latest experiments on the interrelationship between modifying factors, oxygen-dependent damage and the quantity and decay of radiation-induced free radicals. Also, unless stated otherwise, the data will be that obtained from experiments on barley.

Seed Water Content.

For a number of years it was widely accepted that a direct relationship existed between water content and radiosensitivity. An inverse relationship between sensitivity to sparsely ionizing radiations and seed water content (below 13 to 16%) was first demonstrated about 1954.[3,23] Later it was shown that seeds of 4% water content exhibited more damage if they were soaked in oxygen-bubbled water after X-irradiation than if they were soaked in nitrogen-bubbled water.[5] The effect of seed water content on radiosensitivity was extensively studied during the early 1960's.[8,28,33,34,37]

By rigidly controlling atmosphere during and after irradiation, a definite relationship between seed water content and oxygen was established.[12,18] These results are summarized in Table I. The oxygen-enhancement ratio (OER) for seedling injury decreases from ~ 9 to 1 as the seed water content is increased from 2.2 to 13.1%. The OER as used here is a post-irradiation oxygen effect and is defined as the ratio of exposures or doses required to produce the same biological effect in seeds post-treated in oxygen-free and oxygenated water. Thus an OER of 1 indicates no oxygen effect. If oxygen is completely eliminated during and after irradiation, little difference exists in radiosensitivity of seeds between 2 and 13% water; the D_{50} for seedling growth reduction is about 60 kR. If seeds are soaked in oxygen-bubbled water after irradiation, the D_{50} increases from ~ 7 to 60 kR as the seed water content is increased from 2 to 13%. The decrease in the OER with increasing seed water is not a simple linear response. At certain water contents there is very little difference in oxygen sensitivity, whereas at others a small change can greatly alter the response of seeds to oxygen.[15] This is particularly important for seed water contents between 10 and 11.5%. Seeds stored under normal laboratory conditions are likely to be in this range and a difference in water content of as little as 0.2-0.3% might greatly alter radiosensitivity if oxygen availability is not rigidly controlled.[15]

TABLE I. THE INFLUENCE OF SEED WATER CONTENT ON D_{50} VALUES FOR SEEDLING INJURY AND THE OXYGEN-ENHANCEMENT RATIO (OER) IN BARLEY SEEDS RECEIVING VARIOUS EXPOSURES OF ^{60}CO GAMMA RAYS AND THEN SOAKED IN OXYGEN- OR NITROGEN-BUBBLED WATER AT 0° C FOR 18 HOURS. (Summarized from Ref. 12, 18)

Seed water content (%)	D_{50} 1/		OER
	O_2	N_2	
2.2	7.0	62.3	8.9
5.7	11.1	56.3	5.1
7.9	12.3	62.0	5.0
9.7	18.0	63.0	3.5
11.5	43.5	59.3	1.4
13.1	58.3	58.7	1.0

1/ Gamma ray exposure (kR) required to produce 50% seedling injury compared to non-irradiated controls.

Various workers have shown that the quantity of radiation-induced free radicals, measured by EPR, becomes less as the seed water content is increased[8,19,21,32], suggesting a close relationship between radiation-induced free radicals and oxygen-enhanced radiation sensitivity in irradiated seeds.

Post-Irradiation Storage.

Most post-irradiation storage experiments have shown an increase in damage during storage of very dry (~ 4% water) but not moist seeds (≥ 13% water)[20]. These results, however, are for seeds irradiated and stored in air. Later experiments have shown that the increase in damage is probably due to an oxygen effect.[12,18]

The response for seedling injury in barley seeds of different water contents exposed *in vacuo* to 25 kR of ^{60}Co γ-radiation and then stored *in vacuo* for different periods at room temperature before soaking in oxygen- or nitrogen-bubbled water at 0° C is summarized in Fig. 1. If seeds are soaked in nitrogen-bubbled

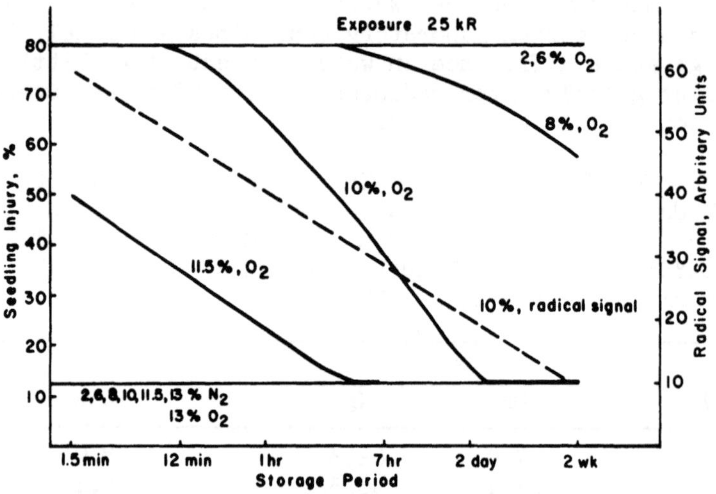

Fig. 1. Relative seedling injury of Himalaya barley seeds of various water contents exposed *in vacuo* to 25 kR of ^{60}Co γ-rays and then stored for various periods *in vacuo* at room temperature before soaking in oxygen or nitrogen-bubbled water at 0° C for 18 hours. Change in radical signal in seeds of 10% water content is also shown. (Summarized from Ref. 12, 18).

water, biological damage is low regardless of seed water content between 2 and 13% and does not change during storage.

Maximal damage occurs in seeds of 2 and 6% water contents if they are soaked in oxygen-bubbled water immediately or at any period ranging up to two weeks after irradiation indicating no loss of oxygen-sensitive sites during storage. There is no oxygen effect in seeds of 13% water content. Damage is minimun immediately after irradiation and after any storage period up to two weeks. This response indicates no available oxygen-sensitive sites even immediately after irradiation. The sites (presumably free radicals) probably disappeared almost instantaneously during irradiation via recombination or some other process.

Seeds of 8, 10 and 11.5% water contents exhibit a loss in oxygen sensitivity during post-irradiation storage. A partial loss in oxygen sensitivity in seeds of 8% water occurs toward the latter part of the storage period suggesting some radical decay. The initial injury level is lower in seeds of 11.5% water content indicating a partial loss of radicals during irradiation. The remainder of the oxygen response disappears within the first few hours of post-irradiation storage.

A maximum loss in oxygen sensitivity during storage occurs in seeds of 10% water content. If the seeds are exposed to oxygen immediately after irradiation damage is maximum but after about two days storage complete loss in oxygen sensitivity occurs and damage is at the minimal level. The loss of oxygen sensitivity in seeds of 10% water is reflected not only in terms of seedling injury but also for frequencies of chromosome aberrations (anaphase bridges and fragments) and chlorophyll deficient mutations as well.[12,18] The loss of oxygen sensitivity is also correlated with a loss in EPR signal during storage, again indicating a relationship between oxygen sensitivity and free radicals.

<u>Temperature</u>.

The influence of temperature on the radiosensitivity of plant seeds has been studied by various workers. In general, they have found that a heat treatment applied at some time before, during or after irradiation has a protective effect.[4,6,25,28,33] Similarly, very low temperatures (-80 to -196° C) provide protection against or retard the development of radiation damage.[4,27]

Recently, experiments were conducted with Himalaya barley seeds to relate the effects of temperature during and after irradiation to oxygen-dependent and -independent damage and free radical decay.[13] In these experiments seed water content was held constant and only temperature varied. Seeds containing

10.0 ± 0.1% water were used since, as shown above, they exhibit the maximum loss of oxygen sensitivity during post-irradiation storage at room temperature. The seeds were irradiated in vacuo in sealed glass vials at temperatures ranging from -78 to 50° C and then soaked either immediately in oxygen- or nitrogen-bubbled water at 0° C for 18 hours or stored at the irradiation temperature for various periods before soaking.

Results of these experiments have been presented in detail[13] and will only be summarized here. The OER for seedling injury is approximately 4 for seeds irradiated at temperatures ranging from -78 to 10° C. The OER decreases with increasing temperature, and at 50° C the OER = 1. Between 30 and 40 C is especially critical; a small change in temperature can greatly alter the response to oxygen. In contrast to a decrease in oxygen-dependent damage with increasing temperature, oxygen-independent damage increases approximately 1.4-fold as the irradiation temperature is increased from - 78 to 50° C.

The influence of temperature on the decrease in oxygen sensitivity and free radicals during post-irradiation storage is summarized in Table II. This table is self-explanatory, but a few points will be emphasized. Although both oxygen sensitivity and free radical signal decrease during post-irradiation storage at certain temperatures, the decreases are not quite parallel. At 50° C there is no oxygen effect immediately after irradiation, yet there is a sizeable radical signal which decreases during storage. Apparently, at least two radical components exist, one which is related to oxygen-dependent damage and one which is not. The oxygen-related component appears to decay at a faster rate than the oxygen-independent one. At - 78° C there is no decay of free radicals, while at 0° C there is some decay of the oxygen-dependent component. At 30° C there is probably complete disappearance of the oxygen-dependent and partial decay of the oxygen-independent component. The radical decay observed during storage at 50° C is probably only oxygen-independent radicals. The increase in damage during storage at 50° C represents an oxygen-independent class which develops very slowly and probably requires several months if seeds are stored at room temperature.[18]

Rates of Development of Oxygen-Dependent Damage and Loss of Oxygen-Sensitive Sites During Soaking.

The induction of a post-irradiation oxygen effect by soaking seeds in oxygen-bubbled water was first demonstrated by Caldecott et al.[5] These workers also stated that after soaking 45 minutes in nitrogen-bubbled water an oxygen-effect could still be induced by transferring the seeds to oxygen-bubbled water. These experiments were extended to include seeds of different water contents, temperatures and EPR measurements.

TABLE II. SUMMARY FOR THE EFFECTS OF TEMPERATURE DURING POST-IRRADIATION STORAGE ON CHANGES IN OXYGEN-SENSITIVITY (MEASURED BY SEEDLING INJURY) AND FREE RADICAL SIGNAL (MEASURED BY EPR) IN BARLEY SEEDS OF 10.0% WATER CONTENT EXPOSED IN VACUO TO 25 kR OF ^{60}Co γ-RADIATION AND THEN STORED IN VACUO FOR VARIOUS PERIODS RANGING UP TO TWO WEEKS BEFORE SOAKING IN OXYGEN- OR NITROGEN-BUBBLED WATER FOR 18 HOURS AT 0° C (Summarized from Ref. 13).

Irradiation and storage temperature	Soaking atmosphere	Change in seedling injury or radical signal
-78 to 50° C	N_2	Injury is minimum immediately after irradiation and does not change during a post-irradiation storage period of 2 weeks, except for a slight increase in damage toward the end of the storage period in seeds stored at 50° C.
-78° C	O_2	Injury is maximum immediately after irradiation and there is no decrease during storage.
0° C	O_2	Injury is maximum immediately after irradiation and decreases slightly toward end of 2-week storage period.
30° C	O_2	Injury is almost maximum immediately after irradiation, but O_2 sensitivity decreases during storage and minimum damage is attained after only 7 hrs storage.
50° C	O_2	Injury is minimal immediately after irradiation. Slow increase in damage (not O_2-dependent) as storage progresses.
-78 to 50° C (EPR)		No change in signal at -78° C; signal 120 arbitrary units immediately after irradiation and at the end of 2 weeks storage. Decrease from 120 to 75 units at 0° C, from 100 to 25 at 30° C and from 55 to 10 at 50° C.

Himalaya barley seeds of 1.2 to 12.8% water content were exposed in vacuo to 25 kR of ^{60}Co γ-radiation and then placed immediately in oxygen- or nitrogen-bubbled water at 0° C. They were allowed to soak for various periods ranging from 5 minutes to 8 hours before transfer to the alternate soaking atmosphere for the remainder of a 16-hour soaking period. It was hypothesized that if irradiated seeds were soaked in oxygen-bubbled water for a period of time a certain amount of the radiation-induced free radicals would react with oxygen. On transfer to oxygen-free water the remainder would recombine. The rate of development of oxygen-dependent damage could then be determined by transferring different samples at different times. Similarly the rate of loss of oxygen-sensitive sites could also be determined by placing seeds initially in oxygen-free water and then transferring them to oxygenated water at various times. This would allow a certain amount of radicals to recombine in the absence of oxygen, the remainder being available to react with oxygen upon transfer to oxygenated water. By transferring different samples at different times the rate of loss of oxygen-sensitive sites could be determined.

Results for seeds of 1.2% water content are shown in Fig. 2. The development of oxygen-dependent damage is almost completed at the end of two hours soaking in oxygen-bubbled water. However, at the end of a two-hour soaking period in oxygen-free water a high degree of oxygen-sensitivity still remains indicating little radical recombination during the first two hours soaking at 0° C. Eight hours soaking in oxygen-free water are required before no oxygen-effect is observed when the seeds are transferred to oxygenated water.

The rates of these reactions were not changed by increasing seed water content. The initial and final injury levels were much lower as the seed water content approached 11.5% but the same periods of at least 2 and 8 hours were required for complete development of oxygen-dependent damage and loss of oxygen-sensitivity, respectively. Seeds of 12.8% water content showed no response to oxygen.[17]

Increasing the soaking temperature increases the rates of both phenomena by approximately 2.5 to 3 times for each 10° C rise in temperature (Fig. 3). Whereas it requires between 2 and 4 hours for the complete development of oxygen-dependent damage and at least 8 hours for complete loss of oxygen-sensitivity at 0° C, these events at 30° C require only about 10 and 30 minutes, respectively. Regardless of the temperature, however, it still requires about 3 times as long for complete loss of oxygen sensitivity than for complete development of oxygen-dependent damage.

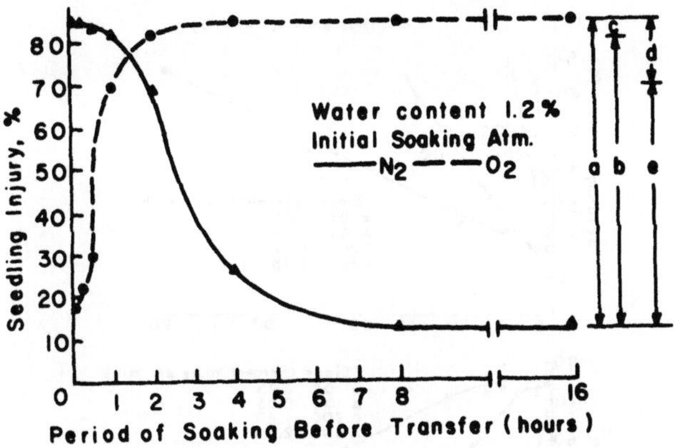

Fig. 2. Development of Oxygen-dependent Damage (dashed line) and Loss of Oxygen-sensitive sites (solid line) in irradiated barley seeds of 1.2% water content. The seeds were exposed <u>in vacuo</u> to 25 kR ^{60}Co gamma rays, then soaked in oxygenated or oxygen-free water at 0° C for various periods of time before transfer to the opposite soaking atmosphere. (a) Total oxygen-dependent damage. (b) Oxygen-dependent damage developed after a 2-hour soaking period in oxygenated water at 0° C. (c) Oxygen-dependent damage remaining after a 2-hour soaking period in oxygenated water at 0° C. (d) Oxygen sensitivity lost after a 2-hour soaking period in oxygen-free water at 0° C. (e) Oxygen sensitivity remaining after a 2-hour soaking period in oxygen-free water at 0° C. (From Ref. 17).

Although not shown here, the loss of free radicals during soaking at 0° C in seeds of 2.1% water content was also measured.[17] Two radical components were observed. One disappeared considerably faster than the development of oxygen-dependent damage and the other very much slower, in fact the second component appeared to be quite resistant to hydration. The pattern was essentially the same regardless of whether the seeds were soaked in oxygenated or oxygen-free water and a sizeable free radical signal remained (about 50% of maximum) after 16 hours soaking. The EPR data have been recently confirmed by Ahnstrom and Sanner.[1]

During post-irradiation soaking it appears that oxygen can move into the seed and react with radiation-induced free radicals at a much faster rate than the radicals can recombine. A possible

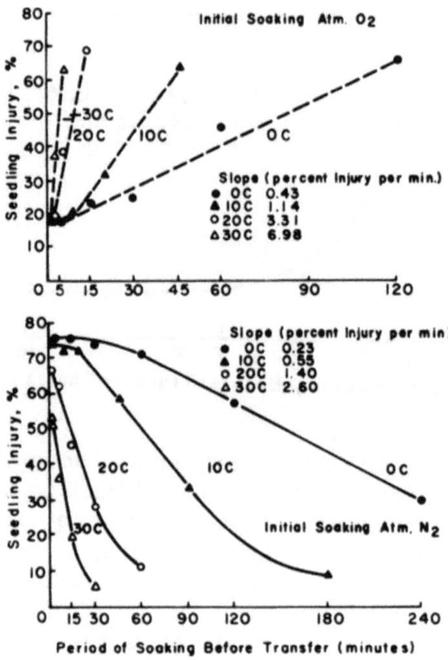

Fig. 3. Influence of soaking temperature on the rates of development of oxygen-dependent damage (top) and loss of oxygen-sensitivity (bottom) as measured by seedling injury in barley seeds of 10.0% water content exposed <u>in vacuo</u> to 25 kR of ^{60}Co γ-radiation.

explanation is that an oxygen molecule is probably quite small compared to the free radical, many which have probably arisen from large organic macromolecule and may even be fixed in the structure of the cell material. Hence, the mobility of oxygen may be less restricted and more likely to reach a position where it can react with the radicals.

The results for radical decay during soaking do not correspond directly with the biological data suggesting that processes (physical or chemical) leading to biological damage may be even more complicated than we first realized. The previous evidence for more than one radical component,[22,29] with at least one component being hydration resistant,[24] was confirmed in these studies. These studies, along with those presented above for post-irradiation storage at different temperatures suggest the definite presence of radiation-induced free radicals which are not associated with oxygen-dependent damage.

Ahnström and Sanner[1] recently reported that the free radicals induced in the embryo and endosperm have distinctly different lifetimes. Those in the embryo disappear much more rapidly during post-irradiation soaking than those induced in the endosperm. In fact, the latter are quite stable. It is generally assumed that radiation damage is localized to certain target structures of cells in the embryo, particularly chromosomes thus, relating oxygen-dependent damage directly to free radicals induced in whole seeds may not be valid, especially since the embryo makes up only a small part of the whole seed. Thus, it seems that free radicals involved in oxygen-enhanced radiation damage more closely correspond to the lifetime of those in the embryo.[1] Since free radicals induced in the endosperm are stable and also probably play a minor role in the induction of damage they may correspond to the oxygen-independent and hydration resistant radicals described above.

Studies with Seeds of Different Plant Species.

The foregoing are results obtained with seeds of a single species, viz., barley. Previous studies have shown that seeds of different species equilibrate for water content at different rates and also differ in radiosensitivity when equilibrated at the same relative humidity.[36,37] Since, as shown above for barley, the main role of water content appears to be that of regulating the oxygen effect, studies were initiated to study the post-irradiation oxygen sensitivity of dry seeds of different species. Also, there has been a tendency to extrapolate results obtained with barley as being general for seeds of all plant species.

Seeds of 6 species, *Hordeum vulgare* (barley), *Oryza sativa* (rice), *Allium cepa* (onion), *Medicago sativa* (alfalfa), *Cucumis sativas* (cucumber), and *Raphanus sativas* (radish), were desiccated to a water content of between 1 and 3%. These species were chosen because they differ widely in radiosensitivity and in nuclear and interphase chromosome volumes.[36] The seeds were then sealed *in vacuo* in glass vials and subjected to varying exposures of ^{60}Co γ-rays. After irradiation they were soaked in oxygen- or nitrogen-bubbled water at 0° C for 18 hours. Seedlings were cultured in a sand-peat-perlite mix in greenhouse benches or in flats in a growth room except that barley seedlings were cultured on blotters. Growth above ground level for the monocotyledonous species and above the cotyledonary nodes of the dicotyledonous species was harvested after an arbitrary period. Oven dry weights of irradiated treatments were compared to those of a non-irradiated control.

The results (Table III) show a wide difference in oxygen sensitivity among the species studied. The D_{50} for seeds soaked in oxygen-bubbled water ranges from 4.5 kR for rice to 59 kR for radish, a difference of more than 13-fold. The D_{50} for seeds soaked in nitrogen-bubbled water ranges from 17.5 kR for onion to 137 kR for radish, a difference of almost 8-fold. The OER ranges from 1.4 for onion to 16.7 for rice, a difference of almost 12-fold. Thus, a wide difference exists among species in their response to oxygen following gamma-irradiation of seeds which is not related to nuclear or interphase chromosome volumes, the relative radiosensitivity of the species or to any taxonomic criteria. For example, radish, a dicot and onion, a monocot, were the species least sensitive to post-irradiation oxygen treatment, yet under anoxic conditions radish was the most radioresistant and onion the most radiosensitive. At present, it appears that the chemical composition of the seeds may be an important factor governing the response of seeds to oxygen following gamma-irradiation. Seeds of the two most sensitive species, barley and rice, are very high in starch compared to the somewhat oily seeds of onion and radish.

TABLE III. D_{50} VALUES (EXPOSURES REQUIRED TO REDUCE SEEDLING GROWTH 50%) AND OER'S OF VERY DRY (< 3% WATER CONTENT) SEEDS OF SIX SPECIES OF HIGHER PLANTS IRRADIATED IN VACUO WITH VARYING EXPOSURES OF ^{60}Co γ-RAYS AND THEN SOAKED IN OXYGEN- OR NITROGEN-BUBBLED WATER 18 HOURS AT 0° C.

Species	$D_{50}O_2$ kR	$D_{50}N_2$ kR	OER
Allium cepa	13.0	17.5	1.4
Raphanus sativus	59.0	137.0	2.3
Medicago sativa	21.5	78.0	3.6
Cucumis sativus	6.3	45.5	7.2
Hordeum vulgare	5.0	47.0	9.4
Oryza sativa	4.5	75.0	16.7

Cervigni et al.[7] have related seed radiosensitivity to thermoluminescence (TL) yields, i.e., the more resistant species have larger glow curves than the sensitive ones. The saponifiable lipid component appears to be most important. These authors suggest that differences in seed radiosensitivity may be due to differences in the species ability to retain energy. The radio-resistant species re-emit more energy while the radio-sensitive species retain the energy. Whether or not these results can be related to oxygen-sensitivity has not been tested. The difference in post-irradiation oxygen-sensitivity is probably related to the quantity and/or quality of the oxygen-sensitive sites induced by the irradiation. There are various possibilities which could exist. The sites induced in starchy seeds may have a stronger affinity for oxygen, have longer lifetimes or the peroxy radicals formed may be more effective in causing biological damage. Singh et al.,[40] working with *Zea mays*, found more radiation-induced free radicals in seeds of low oil content than in seeds of high oil content but the low oil seeds were also more radio-resistant. Since these results are for a single species, further studies in this area are needed to provide more information on the mechanisms by which γ-radiation induces biological damage, especially in dry systems.

The above results also have a practical significance since seeds are the most frequently treated growth stage for the purpose of inducing mutations. Therefore, in order to apply effective radiation doses and obtain meaningful and reproducible results, it is important to know the extent to which modifying factors, such as oxygen, should be controlled when treating seeds of the particular species with which one is working.

CONCLUSIONS

Seeds of higher plants have been extensively used as a radiobiological test system to advance our knowledge on certain mechanisms, particularly the early events involving the production and decay of free radicals, by which ionizing radiations induce damage in living cells. Most of the information gained has come from studies involving different modifying factors.

Oxygen and free radicals appear to be closely related in the induction of at least part of the radiation-induced damage. Seed water content and temperature appear to regulate radical mobility and hence the degree to which free radicals are available to react with oxygen immediately after irradiation or at some time thereafter. By manipulating seed water content and temperature we can control the degree of free radical recombination and hence study the reaction of radicals and oxygen over a specified period of time during post-irradiation storage.

Although it is certain that free radicals in combination with oxygen are responsible for a part of the biological damage caused by high energy photons[38] there may not be a simple 1 to 1 relationship. There appear to be more than one radiation-induced free radical component with at least one being quite stable and hydration-resistant. The stable radicals may be those not associated with oxygen-dependent damage and also induced primarily in the endosperm.

Post-irradiation storage and soaking experiments in seeds of different species coupled with EPR studies might provide some answers to radical lifetimes and decay and also aid in the identification of radical species.

REFERENCES

1. Ahnstrom, G. and Sanner, T. (1971) Effect of hydration on the decay of radiation-induced radicals and oxygen-sensitive centers in barley seeds. Radiat. Bot. 11:27-32.

2. Bacq, Z. M. and Alexander, P. (1961) Fundamentals of Radiobiology. Second Edition. Pergamon Press, London, 555 p.

3. Caldecott, R. S. (1954) Inverse Relationship between the water content of seeds and their sensitivity to X-rays. Science 120:809-810.

4. Caldecott, R. S. (1961) Seedling height, oxygen availability, storage and temperature: Their relation to radiation-induced genetic and seedling injury and barley. p. 3-24. In: Effects of Ionizing Radiations on Seeds. IAEA, Vienna.

5. Caldecott, R. S., Johnson, Elizabeth B, North, D. T. and Konzak, C. F. (1957) Modification of radiation-induced injury by posttreatment with oxygen. Proc. Nat. Acad. Sci. (U. S.) 43:975-983.

6. Caldecott, R. S. and Smith, L. (1952) The influence of heat treatments on the injury and cytogenetic effects of X-rays on barley. Genetics 37:136-157.

7. Cervigni, T., Bassanelli, C., Caserta, G., and Laitano, R. F. (1969) Thermoluminescence in complex biochemicals: A possible correlation with radiosensitivity. Radiat.Res. 38:579-587.

8. Conger, A. D. (1961) Biological after-effect and long-lived free radicals in irradiated seeds. In: Symposium on Recovery of Cells From Injury. J. Cell. Comp. Physiol. 58:(Suppl.1) 27-32.

9. Conger, A. D. (1963) Chromosome aberrations and free radicals. p. 167-202. In: S. Wolff (ed.), Radiation-Induced Chromosome Aberrations. Columbia Univ. Press, New York.

10. Conger, A. D. and Stevenson, H. Q. (1969) A correlation seedling height and chromosomal damage in irradiated barley seeds. Radiat. Bot. 9:1-14.

11. Conger, A. D. and Randolph, M. L. (1959) Magnetic centers (free radicals) produced in cereal embryos by ionizing radiation. Radiat. Res. 11:54-66.

12. Conger, B. V. (1967) The influence of seed water content on the oxygen effect in irradiated barley seeds. Ph. D. Thesis, Washington State University. 91 p. Typed.

13. Conger, B. V., Hileman, J. R., Nilan, R. A. and Konzak, C. F. (1971) The influence of temperature on radiation-induced oxygen-dependent and -independent damage in barley seeds. Radiat. Res. 46:601-612.

14. Conger, B. V., Konzak, C. F. and Harle, J. R. (1969) A glass manifold vacuum system for controlling atmosphere and water content of seeds for irradiation experiments. Radiat. Bot. 9:425-427.

15. Conger, B. V., Nilan, R. A. and Konzak, C. F. (1968) Post-irradiation oxygen sensitivity of seeds varying slightly in water content. Radiat. Bot. 8:31-36.

16. Conger, B. V., Nilan, R. A. and Konzak, C. F. (1968) Radiobiological damage: A new class identified in barley seeds by post-irradiation storage. Science 162:1142-1143.

17. Conger, B. V., Nilan, R. A. and Konzak, C. F. (1969) The role of water content in the decay of radiation-induced oxygen-sensitive sites in barley seeds during post-irradiation hydration. Radiat. Res. 39:45-56.

18. Conger, B. V., Nilan, R. A., Konzak, C. F. and Metter, S. (1966) The influence of seed water content on the oxygen effect in irradiated barley seeds. Radiat. Bot. 6:129-144.

19. Cook, R. F. (1963) The effect of water and a protective agent on gamma-ray induced free radicals in mustard seeds. Int. J. Rad. Biol. 7:497-504.

20. Curtis, H. J., Delihas, N., Caldecott, R. S. and Konzak, C. F. (1958) Modification of radiation damage in dormant seeds by storage. Radiat. Res. 8:526-534.

21. Ehrenberg, A. (1961) Research on free radicals in enzyme chemistry and in radiation biology. p. 337-350. In: M. S. Blois Jr. (ed.), Free Radicals in Biological Systems. Academic Press, New York.

22. Ehrenberg, A. and Ehrenberg, L. (1958) The decay of X-ray induced free radicals in plant seeds and starch. Arkiv Fysik 14:133-141.

23. Ehrenberg, L. and Nybom, N. (1954) Ion density and biological effectiveness of radiation. Acta. Agr. Scand. 4:396-418.

24. Haber, A. H. and Randolph, M. L. (1967) Gamma-ray-induced ESR signals in lettuce: Evidence for seed-hydration-resistant and -sensitive free radicals. Radiat. Bot. 7:17-28.

25. Kempton, J. H. and Maxwell, L. R. (1941) Effect of temperature during irradiation on the X-ray sensitivity of maize seed. J. Agr. Res. 62:603-618.

26. Konzak, C. F., Bottino, P. J., Nilan, R. A. and Conger, B. V. (1968) Irradiation of seeds: A review of procedures employed at Washington State University. p. 83-96 In: Neutron Irradiation of Seeds II. IAEA, Vienna.

27. Konzak, C. F., Curtis, H. J., Delihas, N. and Nilan, R. A. (1960) Modification of radiation-induced damage in barley seeds by thermal energy. Can. J. Genet. Cytol. 2:129-141.

28. Konzak, C. F., Nilan, R. A., Harle, J. R., and Heiner, R.E. (1961) Control of factors affecting the response of plants to mutagens. Brookhaven Symp. Biol. 14:128-157.

29. Löfroth, G., Ehrenberg, A. and Ehrenberg, L. (1964) Analysis of Radiation induced electron spin resonance spectra in plant seeds. Radiat. Bot. 4:455-467.

30. Lunden, A. O. (1964) Seed embryo features and radiation response. Radiat. Bot. 4:429-437.

31. Myhill, R. R. and Konzak, C. F. (1967) A new technique for culturing and measuring barley seedlings. Crop Sci. 7:275-277.

32. Myttenaere, C., Bourdeau, Ph., Helcke,G. and Masset, M. (1965) Radiosensitivity of rice seed in relation to water content and free radicals. Radiat. Bot. 5:443-451.

33. Nilan, R. A., Konzak, C. F., Harle, J. R. and Heiner, R. E. (1962) Interrelation of oxygen, water and temperature in the production of radiation-induced genetic effects in plants. Strahlentherapie (Suppl.) 51:171-182.

34. Nilan, R. A., Konzak, C. F., Legault, R. R. and Harle, J. R. (1961) The oxygen effect in barley seeds. p. 139-154. In: Effects of Ionizing Radiations on Seeds. IAEA, Vienna

35. Nilan, R. A., Konzak, C. F., Wagner, J., and Legault, R. R. (1965) Effectiveness and efficiency of radiations for inducing genetic and cytogenetic changes. In: The Use of Induced Mutations in Plant Breeding. Radiat. Bot. 5:(Suppl.) 71-89.

36. Osborne, T. S. and Lunden, A. O. (1965) Prediction of seed radiosensitivity from embryo structure. In: The Use of Induced Mutations in Plant Breeding. Radiat. Bot. 5(Suppl.) 133-149.

37. Osborne, T. S., Lunden, A. O. and Constantin, M. J. (1963) Radiosensitivity of Seeds III. Effect of pre-irradiation humidity and gamma-ray dose on seeds from five botannical families. Radiat. Bot. 3:19-28.

38. Powers, E. L. (1966) Contributions of electron paramagnetic resonance techniques to the understanding of radiation biology. pp. 137-159. In: Electron Spin Resonance and the Effects of Radiation on Biological Systems. NAS-NRC, pub. 1355, Washington, D. C.

39. Powers, E. L. and Tallentire, A. (1968) The roles of water in the cellular effects of ionizing radiations. p. 1-67 In: Actions Chimiques et Biologiques des Radiations. Mason et Cie, Paris.

40. Singh, B. B., Venkaterman, B., Notani, N. K. Mouli, Chandra and Bora, K. C. (1962) Role of chemical composition of biological material on the production and decay of radiation-induced free radicals and their identification. p. 2-13. In: Symposium on the Biological Effects of Ionizing Radiation at the Molecular Level. IAEA, Vianna.

41. Zimmer, K. G., Ehrenberg, L. and Ehrenberg, A (1957) Nachweis langlebiger magnetischer Zentren in bestrahlen biologischen Medien and deren Bedentung für die Strahlenbiologie. Strahlentherapie 103:3-15.

-DISCUSSION-

DAVIDSON - Is there experimental evidence for the presence of free radicals in embryos alone as opposed to whole seeds?

CONGER, A. - Yes, there are free radicals in the embryos.

DAVIDSON - What about ESR signals in embryos?

CONGER, A. - ESR signals have been studied in the different biochemical fractions of embryos, but not as far as I'm aware in different parts of seed embryos. These fractions induced an oily fraction, nuclear fraction, cell wall fraction, etc. There was no signal from the oily fraction.

CONGER, B. - The absence in a signal from the oily fraction may explain why the post-irradiation oxygen-enhanced damage is less in seeds which are high in oil content.

D'AMATO - We have undertaken embryo transplantation experiments where the embryos of wet *Durum* wheat seeds are excised and then transplanted to another endosperm (with about 95 percent efficiency). When unirradiated embryos are transplanted to irradiated endosperms we observed chromosome aberrations in the root and chlorophyll mutations in the spike progeny. This response is X-ray energy dependent; 140-160 kV -- but not 220 kV -- induced the aberrations and mutations. Our interpretation is that we induced metabolic derangements which acted as chemical mutagens. That all the chromosome aberrations were of the chromatid-type supports this interpretation.

BALL - There may be some significance for seed embryo transplantation in that the outermost layer of the embryo, the aleurone -- which is very much alive -- is placed next to the endosperm which essentially is a dead tissue.

KONZAK - I think it could be very useful to have rate of development responses -- i.e., a series of curves -- on oxygen enhancement ratios for these seed systems.

CONGER, B. - I agree, but this would be very difficult to achieve because the groups would involve large numbers of individuals and there is difficulty in achieving uniform seedling growth amongst all individuals for some of the species with which we have worked.

RADIATION PROBES OF MAIZE MERISTEMS*

H. H. Smith

Biology Department, Brookhaven National Laboratory

Upton, New York 11973

INTRODUCTION

This paper reviews ways that irradiation of dry seeds in maize can be used to gain information about the stem apex and leaf primordia. For the most part the experiments were originally designed to investigate the biological effectiveness of different kinds of radiation; however, they are discussed here as a means of elucidating properties of meristems. Deuteron microbeam experiments were carried out to study the effect of irradiating only specific parts of the dormant embryo, rather than the whole seed. This was accomplished with a deuteron beam of sufficiently narrow width to localize the irradiation.[16] As a consequence, a particular region, primordium or meristematic area could be marked in the embryo by the damage produced or the mutations induced, then traced by effects observed in the developed seedling. The purpose of reporting experiments on comparative responses to x rays vs. neutrons in the context of this conference is to call attention to the consistent and quantitatively predictable performance of the meristem during recovery from various levels of genetic damage inflicted by two different radiation probes of the seed embryo. Observations showing different radiation sensitivities for different embryonic leaves were used to analyze differences in size of nuclei and constituent chromosomal segments in leaf primordial cells.

---

*Research carried out at Brookhaven National Laboratory under the auspices of the U. S. Atomic Energy Commission.

MATERIAL

In seeds of the strain of maize (Zea mays L.) used, primordia of the first 6 leaves only are present in successively earlier stages of development (Fig. 1). The material is of a genetic stock heterozygous for the yellow-green locus (Yg_2/yg_2). The recessive gene, yg_2, gives a yellow-green color in leaves when homozygous or hemizygous, and the presence of the dominant gene, Yg_2, produces full green color, so that the heterozygote has normal green leaves. Loss of the Yg_2 allele (a specific locus deletion), or change in its function (mutation) in heterozygotes gives a yellowish green phenotype in leaf cells in which the phenomenon occurs and in cell lineages of the altered genotypes.[12]

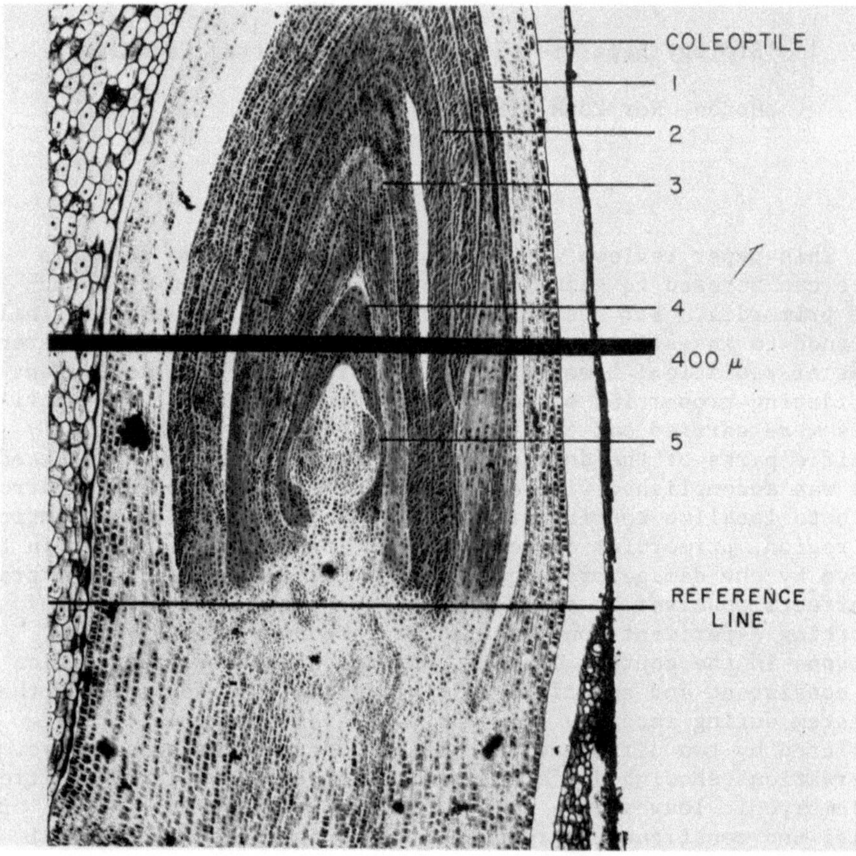

Fig. 1. Longitudinal section of stem apex region of maize embryo, x 125, showing stages of development of the first 6 leaves. The black line 400 μ above the reference line is drawn to scale, 25 μ in width, to represent a microbeam (from Ref.16, by permission of Pergamon Press).

METHODS AND RESULTS

Deuteron Microbeam Experiments

Before irradiation moisture level in the seeds was equilibrated in an atmosphere of 52% relative humidity to about 10.5%. Following irradiation all seeds were hydrated simultaneously on moist absorbant paper at 70° F in the dark. After 7 days the seedlings were planted in soil in pots and grown in the greenhouse at temperatures of 65° F during the day, extended to 18 hr by supplementary lighting, and 60° F at night. Effects of the radiations were scored on the first few seedling leaves.

Irradiations were performed with a beam from the 60-inch cyclotron of Brookhaven National Laboratory (Fig. 2) which produces deuterons with an energy of 22.5 MeV.[1] A desired beam width of 25 μ was achieved by means of a narrow slit, at the exit port of the beam pipe, made by shielding with 0.4 mm platinum foil which is just thick enough to stop the deuterons. Radiation doses of from 35 to 420 krad were used. Approximately 90% of the radiation is confined within the region of aperture width. The

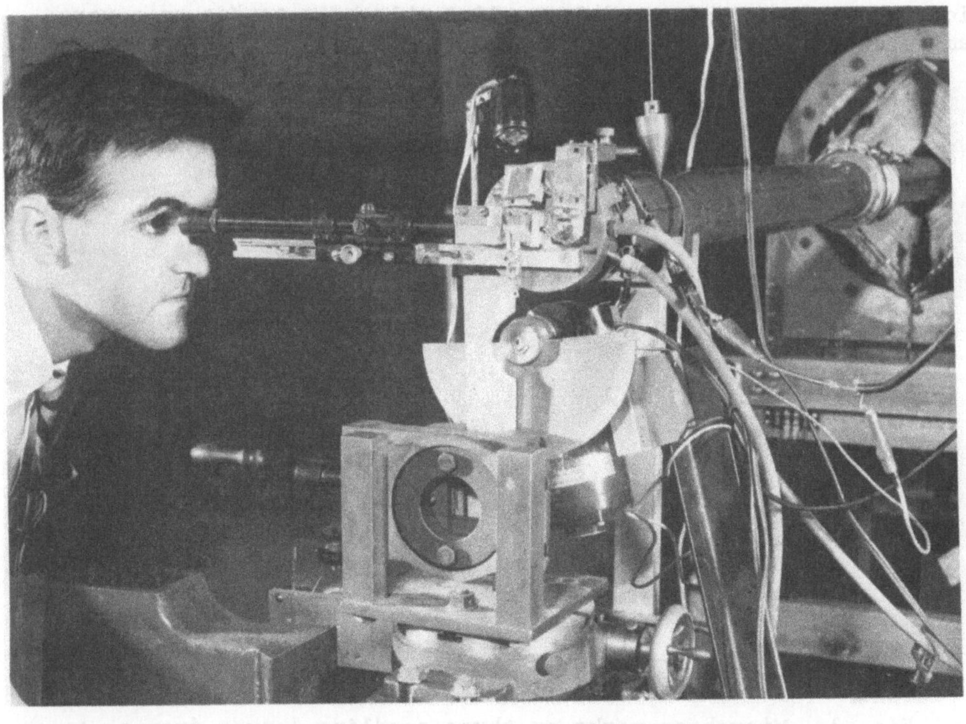

Fig. 2. Arrangement of apparatus for deuteron microbeam irradiation.

22.5 MeV deuteron beam will penetrate about 2 mm of tissue. More energetic cyclotron beams are now available at Brookhaven National Laboratory which will penetrate much deeper, 10 mm for example; however precision decreases with greater penetration due to more scattering of radiation in the tissues.

Accurate alignment of the target region was accomplished by setting cross hairs of a telescope attached to the apparatus in line with a reference point on the embryo side of the seed. From this a reference line was inscribed on the kernel which in selected uniform seeds corresponded in the embryo to the basal region of the stem apex at approximately the base of the first two leaves (Fig. 1). The microbeam was directed to areas of the embryo at specific measured distances above or below the reference line.

The most striking effects on the young plant from microbeam irradiation of the seeds were chlorotic marks that appeared on the first 3 or 4 leaves (Fig. 3). The higher the radiation dose, the more pronounced were the marks. A convenient dose level, for clear marking with little or no detriment to growth, was 88.9 to 125 krad. A cursory examination of tissue in the chlorotic regions, which result from a primary effect of the beam of radiation, shows fewer and possibly smaller chloroplasts but no severe damage to the cells.

Fig. 3. Microbeam marks on three seedling leaves from a single exposure.

The relative positions of the microbeam-induced radiation marks on the seedling leaves can be used: 1) to assess the accuracy of aiming the microbeam at specific areas of the embryo; 2) to locate areas in the mature leaf relative to their position in the embryo; 3) to compare the relative growth in different regions of different leaves from an embryonic to a mature stage; 4) to determine relative radiosensitivities of different specific regions of the seed embryo and to be able experimentally to damage them selectively.

Table I shows the results of aiming a microbeam at three different regions of the embryo: 800, 400 and 325 μ above the reference line. A comparison of these results with the longitudinal section of the stem apex region (Fig. 1) shows that the aim of the beam was fairly accurate in that at 800 μ above the line (toward the top of Fig. 1) the microbeam would be expected to hit near the apex of leaf 1, at the tip of leaf 2 and to miss leaf 3. The marks in fact appeared at 17% of the distance down in leaf 1, 7% in leaf 2, and leaf 3 was unmarked. When the beam was aimed at 400 μ above the line the expected results were again obtained in that the marks appeared near the center of leaves 1 and 2, and about a fifth of the distance down from the tip of leaf 3. When aimed still lower, at 325 μ above the line, the marks appeared more basal (83 and 82%) on leaves 1 and 2, and about 60% of the distance from the tip of leaf 3.

TABLE I

Relation Between Aim of Microbeam and Position of Radiation Mark on Seedling Leaves

Aim of beam μ above reference line	No. seeds irrad.	% leaf length from tip to mark: Leaf 1 $\bar{x}\pm$SE	Leaf 2 $\bar{x}\pm$SE	Leaf 3 $\bar{x}\pm$SE
800	5	17±2.9	7±2.8	0
400	17	57±4.3	52±4.8	22±4.4
325	7	83±2.9	82±2.9	61±4.4

In addition to demonstrating the accuracy and efficacy of the beam in marking specific areas of the embryo in dry seeds, the results also show that the microbeam can be useful in determining with precision the regions of each leaf that were longitudinally contiguous in primordia, though widely separated at maturity.

In subsequent experiments, 7 seedlings were obtained in which each of the first 4 leaves was marked in a single microbeam exposure. The chlorotic lines appeared at the following average distances from the tip: 81% in leaf 1, 80% in leaf 2, 48% in leaf 3, and 5% in leaf 4. These irradiations were made at doses below 210 krad and, though the beam must have passed through or close to the stem apex, the meristem was undamaged and a full complement of 15 leaves developed at maturity. In 5 other seedlings, which were exposed to higher doses of 210 to 420 krad, the beam, in passing through approximately the same region, destroyed the stem apex; and on an average only the 6 leaves, whose primordia were already formed in the embryo, developed. These results suggest that radiation probes of varying dose level could be refined to be still more selective in differentially destroying different parts of the stem or leaf meristems for experimental purposes.

Multiple microbeam exposures of single seeds were made to explore further uses of this tool for botanical research. Each seed was irradiated (88.9 krad) at four different positions spaced 200 μ apart: i.e., 200, 400, 600 and 800 μ above the reference line. The appearance of the leaf markings at maturity is shown in Fig. 4. The distance between marks, set at 200 μ in

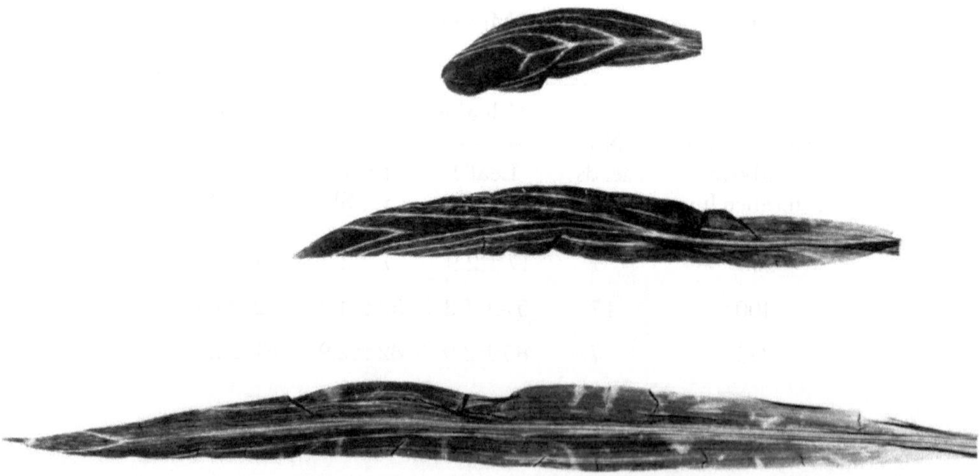

Fig. 4. Microbeam marks on seedling leaves 1, 2 and 3 from four successive exposures made 200 μ apart in the seed (from Ref. 16 by permission of Pergamon Press).

the seed embryo, had now expanded to different distances, depending on the amount of growth that took place in each region of the leaf. The results, from measuring these distances in the mature seedling leaves, are shown in Table II. In each leaf there was considerably more expansion in the basal region than toward the tip. For example, in the upper 10% of leaf 3, marks that were 200 μ apart when made expanded to a distance of 14.5 mm which is a 72-fold increase. Whereas, in the basal 10% of leaf 3 there had been a 385-fold expansion, from 200 μ to 77 mm. The overall linear expansion of leaf 3, from seed embryo to mature seedling, estimated by averaging the regional measurements, was 244-fold, compared to 140-fold in leaf 2 and 42-fold in leaf 1.

TABLE II

Relative Growth of Leaves as Measured by Microbeam Marks Made at 200 μ Intervals in the Seed Embryo

Position of marked interval in % distance from leaf tip	Leaf 1 mm between marks	Leaf 1 Growth increment	Leaf 2 mm between marks	Leaf 2 Growth increment	Leaf 3 mm between marks	Leaf 3 Growth increment
10					14.5	72
20			14.3	72	20.8	104
30	6.0	30	20.0	100		
40	8.0	40	21.7	108	58.0	290
50	7.0	35	30.0	150		
60	10.7	54	32.7	164	93.7	468
70	9.8	49	31.5	158		
80						
90	13.5	68	41.3	206		
100	12.5	62	38.0	195	77.0	385
Average	8.3	42	28.0	140	48.0	244

The deuteron microbeam thus provides a potentially useful tool for making accurate measurements of relative growth in developmental studies tracing seed meristems to mature plant organs.

Comparative Response to X Rays vs. Neutrons

Dormant seeds of the heterozygous $\underline{Yg_2}/\underline{yg_2}$ stock were equilibrated in an atmosphere of 35% relative humidity to a water content of 6.7% and a hydrogen content of 8.1% in the embryo. They were irradiated with 250-kVp x rays (30 ma, 24 cm, 1-mm Al filter, HVL = 4.5 mm Al, dose rate 1800 R/min) or with fission (fast) neutrons generated by thermal neutron bombardment of a ^{235}U plate in the thermal column of the Brookhaven Graphite Reactor (dose rate, 170 rad/min). The neutron energies were those of a slightly degraded fission spectrum with a peak of about 1 MeV and 10%, by physical dosimetry, gamma-ray contamination.[5]

The dose average rate of linear energy transfer (LET) of the neutrons has been estimated as 49 KeV/μ[5] and that of the x rays as about 2 KeV/μ. The two kinds of radiation constitute, therefore, two different probes, one of relatively densely ionizing paths from neutrons and one of relatively sparsely ionizing paths from x rays.

Plants raised from irradiated and from control lots of seeds were generally replicated at least three times in randomized blocks of 10 per treatment, and were grown in chambers maintained at 75° F and 18 hr daylength. For observation of late effects on reproductive characteristics the plants were transplanted into the field and grown to maturity.

Nine different criteria were used to evaluate the effects of irradiation. They are described below in chronological order as measured during plant development subsequent to irradiation and hydration of the seeds.

1) In 5 days the maize seedlings began to emerge above ground. The total number to appear for each treatment was noted (up to 30 days), and the dose required to reduce emergence by 50% was computed from least squares regression lines fitting the linear portion of the dose-response range.[15]

2) Also computed from these same data was the dose at which fewest seedlings emerge and after which there is a "reversal" effect to give more emergent seedlings.[15]

3) Ten days after sowing, the height of each plant was measured and the dose required to produce a 50% reduction in plant height computed.

4) Also computed from these same data was the dose at which plant height is minimal and after which there is a "reversal" effect to give taller seedlings at still higher doses.[11]

5) At 30 days after sowing, the number of seedlings still alive at each dose was counted and the dose required to reduce survival by 50% was computed.

6) and 7) The fifth seedling leaf reached maturity 35 to 40 days after sowing. The number of $\underline{yg}_2$ sectors per leaf was counted, a dose-response curve plotted, and the dose required to induce an average of one, as well as of two, $\underline{yg}_2$ sectors per leaf was computed.

8) At flowering time, about 60 days after sowing, pollen grains were scored for fertility and the dose required to induce 50% pollen sterility was computed.

9) At the end of the growing season, about 90 days after sowing, the number of kernels on each self-pollinated ear was recorded; and the dose required to produce a 50% reduction in the number of seeds per cob was computed.

In time sequence, end points were scored from 5 to 90 days after irradiation. The results are shown in Table III. The severity of radiation damage ranged from sublethal genetic effects ($\underline{yg}_2$ sectors), to gamete or young embryo lethality (pollen sterility and reduced seed set), to loss of somatic cell reproductive integrity (reduction in plant height and survival), to complete cessation of cell division ("reversal" of emergence and plant height).

TABLE III

Criteria of Response, Doses, and Relative Biological Effectiveness (RBE) for Maize Irradiated With X Rays and Fission Neutrons; Arranged in Chronological Order of Scoring

Days from irradiation to scoring		Criteria	Dose: X rays (Krad)	Dose: Neutrons (Krad)	RBE X/N
1)	5–30	50% emergence	67	6.8	10
2)	5–30	Emergence reversal	93	18	5
3)	10	50% plant height	30	2.5	12
4)	10	Plant height reversal	85	24	3.5
5)	30	50% survival	34	3.3	10
6)	35–40	1 $\underline{yg}_2$ sector per leaf 5	5.2	0.07	76
7)	35–40	2 $\underline{yg}_2$ sectors per leaf 5	9.9	0.27	37
8)	60	50% pollen fertility	38	1.7	22
9)	90	50% seed set	21	0.63	33

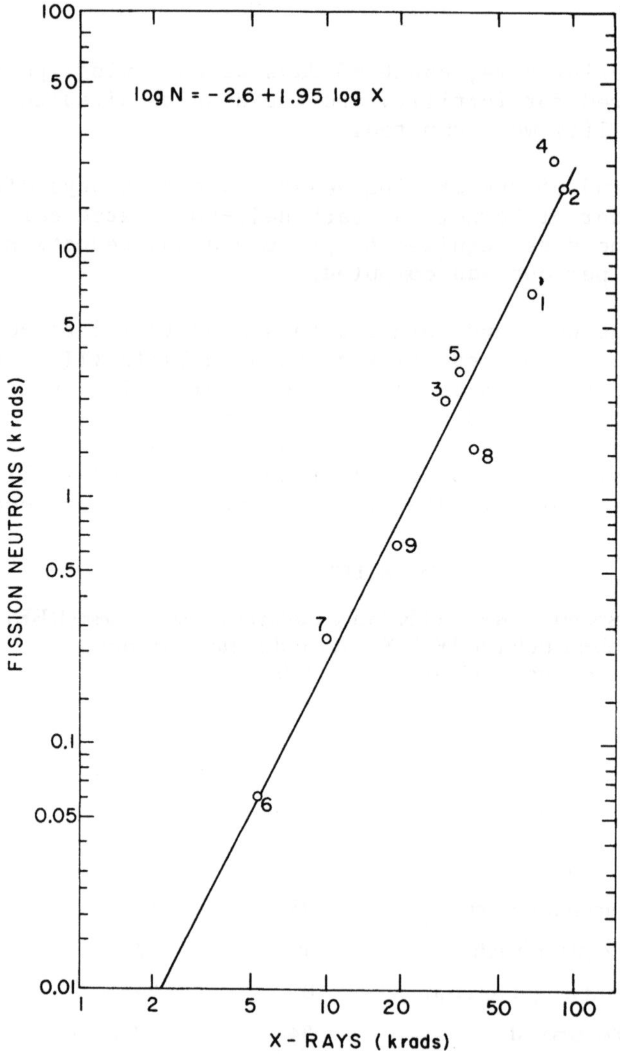

Fig. 5. Logarithmic plot of nine criteria of equal effect (see Table III for reference numbers) in maize plants grown from seed which had absorbed a wide range of doses of fission neutrons or of 250-kVp x rays (from Ref. 15 by permission of IAEA).

A cursory examination of Table III revealed no quantitatively consistent relation between radiation applied and end points scored except that: 1) neutrons (N) were more effective per absorbed dose than were x rays (X); and 2) RBE (X/N dose to give the same effect) was lower the higher the dose required to produce the level of damage scored. For example, neutrons were 76 times more effective per rad than x rays in inducing an average of one $\underline{yg}_2$ sector on leaf 5, but only 3.5 times more effective in inducing plant height "reversal". A marked consistency in these data was made evident when a logarithmic scale plot was constructed with neutron dose (N) on the ordinate and x-ray dose (X) on the abscissa (Fig. 5). The points for equal effect (see Table III) could now be fitted to a straight line (over four orders of magnitude in dose), for which the least squares equation is:

$$\log N = -2.6 + 1.95 \log X$$

That is, the relationship of N to X for equal effect could be expressed as a simple power function (slope = 1.95) over a wide range of doses. These results may be explained on the basis of the assumptions: 1) that neutron induced chromosome breaks and losses increase linearly with dose (e.g., αN); 2) that x-ray induced chromosome breaks and losses increase as the square of the dose (e.g., βX^2). It then follows that $\alpha N = \beta X^2$ or $\log N = \log \beta/\alpha + 2 \log X$. The observed value for the slope of 1.95 is in close agreement with the theoretical expectation of 2 if there is a large two-hit component for x rays.

The relation between genetic damage and cell lethality can be illustrated in another way by an experiment in which the frequency of $\underline{yg}_2$ sectors on leaf 5 was scored over a wide range of neutron and x-ray doses.[18] The results plotted arithmetically, are shown in Fig. 6. The neutron and x-ray curves have in common an initial slope that reaches a peak, saturates, then declines. The curves differ in that: 1) neutrons are much more effective per unit of absorbed dose; 2) the neutron-induced yield of mutations peaks at a lower frequency; and 3) the initial slope of the neutron curve is approximately a straight line whereas with x rays the slope appears to be slightly curvilinear.

The significant relationships can be shown more clearly by plotting both dose and mutation frequency on a logarithmic scale (Fig. 7). Here the initial slope in the neutron curve is drawn arbitrarily as equal to 1 and in the x-ray curve as 2. Because the data fitted these curves fairly well they substantiated the interpretation of: 1) a direct linear relation of neutron dose to number of $\underline{yg}_2$ sectors implying single hit kinetics; and 2) a $(\text{dose})^2$ relationship with x irradiation implying a large two-hit component. The subsequent decrease in the observed number of $\underline{yg}_2$

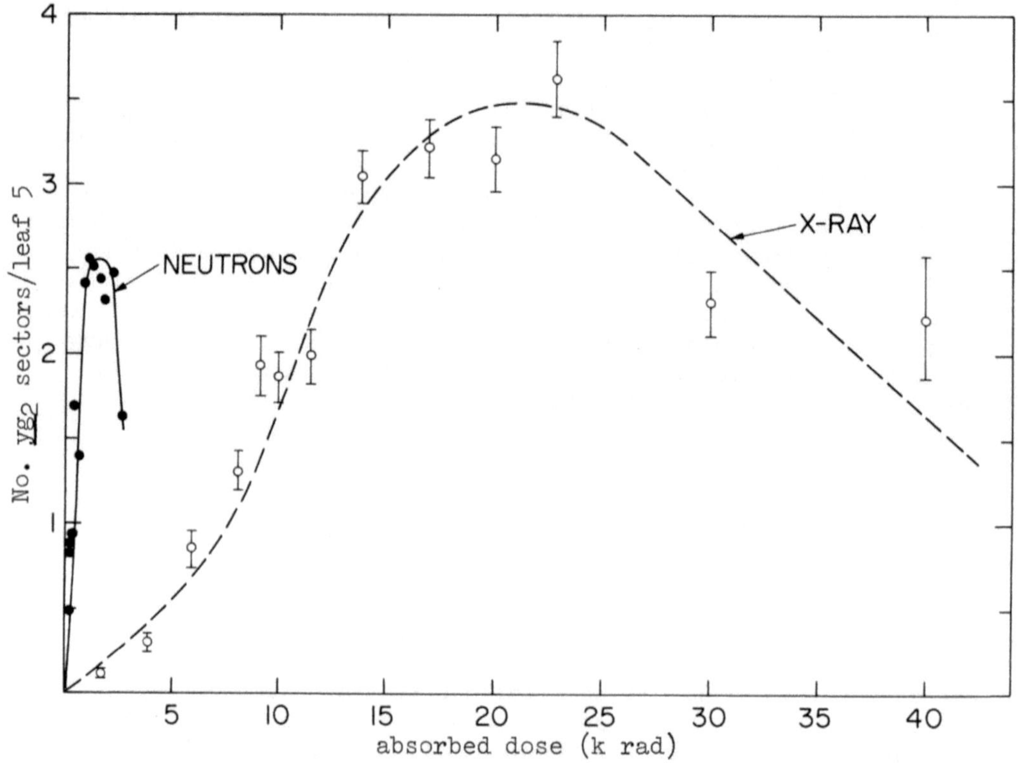

Fig. 6. Frequency of _yg_$_2$ sectors induced in leaf 5 by a wide range of doses of neutrons and x rays; arithmetic plot.

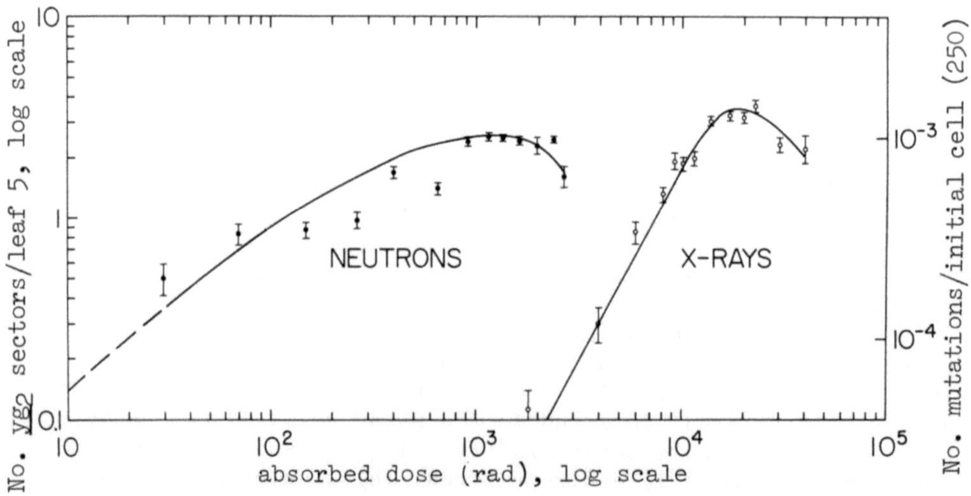

Fig. 7. Frequency of _yg_$_2$ sectors induced in leaf 5 by a wide range of doses of neutrons and x rays; logarithmic plot.

sectors can be attributed to cell killing which is considered to be due to multiple chromosome breaks and losses. This is basically the same phenomenon which causes a $\underline{yg}_2$ sector when it occurs only in the region between the $\underline{Yg}_2$ locus and the centromere with subsequent loss of that particular chromosome segment carrying $\underline{Yg}_2$ in the $\underline{Yg}_2/\underline{yg}_2$ heterozygote.

An effort is being made[19] to fit these experimental results to a mathematical model which is consistent with the following considerations: 1) Both $\underline{yg}_2$-induction and cell killing by radiation are due to chromosome breakage and loss of genetic information. 2) The probability of a break in a chromosome is approximately proportional to its relative length. 3) The number of chromosome breaks increases linearly with neutron dose and with approximately the square of the x-ray dose. It also tentatively appears that: 4) Two breaks on an average, and subsequent loss of the chromosome segments involved, are sufficient to kill a cell; and 5) neutrons can produce multiple breaks along a single ionizing path thus killing the cell in a single event (hence lowering the peak of the neutron curve).

Nuclear Size Differences Among Leaf Primordia

In early experiments[14] on the induction of $\underline{yg}_2$ sectors in maize leaves grown from irradiated seeds, it was shown that the slope of the dose-response curves was different for different leaves. The slopes for linear plots of $\underline{yg}_2$ sector frequency with increasing doses of 250-kVp x rays were: leaf 3 = +0.0011, leaf 4 = +0.0008, leaf 5 = +0.0001. Leaves 3, 4, and 5 were progressively less sensitive to the radiation.

More detailed experiments on $\underline{yg}_2$ induction in leaves 4 and 5 were later made[17] using monoenergetic neutrons as well as x rays. The monoenergetic neutron source and method of irradiation were described in Smith et al.[14] To interpret the $\underline{yg}_2$ frequency, expressed as sectors per leaf, in terms of a mutation rate expressed as breaks per cell, the average number of cells in the primordia of leaves 4 and 5 was determined. In these stocks the leaf 5 primordium appeared in the dormant seed embryo mainly as a four-layered structure with slightly more than 500 cells. However, since the epidermal layers of the leaf are devoid of chloroplasts, and therefore not involved in $\underline{yg}_2$ expression, leaf 5 at the time of irradiation presented a relevant target of about 250 cells. The number of cells in embryonic leaf 4 was more difficult to establish but was estimated by two different techniques to be about 2900. It was concluded that the relevant number of leaf 4 target cells for this experiment was about 1500. This estimate may ultimately be shown to be too high, however, since some of the cells in leaf 4 are differentiated and may not be capable of contributing $\underline{yg}_2$ sectors.

The computed mutation rate and relative biological effectiveness for monoenergetic neutrons of four different energies are shown in Table IV compared to 250-kVp x rays. The results show that the mutation rate per cell is higher in leaf 4 than in leaf 5, that the neutrons (depending on the energy) are 50 to 100 times more effective per rad than the x rays, and that there are certain apparent inconsistencies in the data as, for example, the reversed ranking of RBE's (last two columns in Table IV) for 1.80 vs. 14.7 MeV neutrons in leaves 4 and 5. The data were analyzed in terms of microdosimetric concepts by utilizing detailed physical measurements at the neutron energies employed.[4] One advantage of these experimental results for such analyses is the large differences found in the magnitude of RBE. Also, in considering relative RBE's any errors of dosimetry for the two leaves are ruled out, since the leaves are derived from primordial cells of the same embryo which are separated by but a fraction of a millimeter at the time of irradiation (Fig. 1).

TABLE IV

"Mutation" Rates and RBE's for $\underline{yg}_2$ -Yielding Phenomena in Maize Irradiated With Monoenergetic Neutrons or X Rays

Neutron energy (MeV)	Mutations/rad/cell × 10^5		Relative Biological Effectiveness	
	Leaf 4	Leaf 5	Leaf 4	Leaf 5
0.43	16.8	6.8	102	82
1.25	12.9	5.3	79	64
1.80	10.8	3.9	66	47
14.7	7.7	4.2	47	51
X rays, 250 kVp	0.16	0.08	1	1

Microdosimetric principles show that irradiations with different monoenergetic neutrons produce ionizing paths (individual "events") within spherical microvolumes (targets) that result in the deposition of different amounts of energy with different frequencies.[9] The "event frequency integral" curves generated differ according to the size of the target. Nevertheless, the energy absorbed to produce a mutation should be the same regardless of the different energy deposition patterns or target sizes. Therefore, in order to equalize the energy deposition for the experimentally determined mutation rates, it was necessary to postulate that the target in leaf 4 primordia was larger than in leaf 5 primordia.[17] More specifically, the analysis required the target in leaf 4 to have a diameter of 1.35 μ and in leaf 5, 1.1 μ. Finally, from these computations on target size and energy deposition the seemingly erratic results in Table IV could be accounted for both relatively and absolutely.[17]

The results reported in the first paragraph of this section--that leaves 3, 4 and 5 differ in radiosensitivity--also seemed likely to be explainable as due to differences in target size. Since the volume of the nucleus or average interphase chromosome is an important parameter for whole plant radiosensitivity,[20] it might also be a significant factor in organ radiosensitivity.

To determine the volume of the nucleus at the time of irradiation, primordia were dissected from the seed embryo, placed in a 0.1% solution of the surface-active compound Tween 20 for 5 minutes, then fixed and stained with acetocarmine. The diameter of the nucleus (or length and width, if cylindrical) was measured, and the volume was calculated in cubic microns. The mean nuclear volumes were 207 μ^3 for leaf 3, 159 μ^3 for leaf 4 and 122 μ^3 for leaf 5. There is a direct correlation between nuclear volume and radiosensitivity of these three leaves.

An effort was made to compare estimates of the target size involved in producing a $\underline{yg}_2$ mutation by the method of microdosimetry contrasted to that of measuring nuclear diameter and computing chromosome segment volume and length. The latter estimate was made by assuming the target to be the diameter of that fraction of the total nuclear volume represented by the proportion of the genome that comprises the distance of the $\underline{Yg}_2$ locus from the centromere.[8] The microdosimetric estimate was made by computing the diameter of the assumed spherical target volume and equating this to the $\underline{Yg}_2$-to-centromere distance. The results are shown in Table V. The two estimates are similar. That based on cytological measurement is larger, by 10 to 20% in each leaf, which may be due to: 1) swelling of the nucleus in the course of cytological preparation; or 2) the chromosomes do not make up the full volume of the nucleus.

TABLE V

Estimate of the Distance From $\underline{Yg_2}$ Locus to Centromere

Based on Two Independent Methods of Computation

Computed by: —	Leaf 4 μ	Leaf 5 μ
Nuclear diameter	1.52	1.38
Microdosimetry	1.35	1.10
	0.17	0.28

DISCUSSION

What do the results on comparative effects of x rays and neutrons tell us about the meristem? The primary damage inflicted by each of the irradiations occurred in the seed embryo cells essentially at the moment of exposure. By scoring the effects beginning as early as 5 days after irradiation (emergence) a relationship was established between the dose of neutrons and of x rays to cause the same degree of response. Thereafter, by scoring various other effects and continuing observations until maturity at 90 days after irradiation (seed set), it was found that for each of these end points the relationship of N to X could be described by a single expression--i.e., the equation $\log N = -2.6 + 1.95 \log X$. Moreover, since one of the responses, $\underline{yg_2}$ sectoring, is due largely, if not exclusively, to chromosome breakage and loss, this suggests that all the effects may be attributed mainly to genetic damage[13] and that this increases linearly with the neutron dose and approximately with the square (slope = 1.95) of the x-ray dose.

The main point to be stressed within the context of this conference is that, just as the normal undamaged genotype gives rise to a predictably normal plant development from meristems in the seed embryo, so also does a precisely damaged (in a quantitative sense) genotype in meristematic tissue give rise to an equally predictable series of altered developmental products. Plant development is so specifically programmed by the genetic determinants in the seed meristems, and subsequent development proceeds so specifically under genetic direction, that any quantitative alteration, or "reprogramming", of the genetic information in the primordial cells brings about a quantitatively predictable result at any later stage of growth. Furthermore, the constant dose:(dose)^2 relationship of neutrons to x rays implies a common target for all the measured effects, which follows single-hit kinetics with neutrons and two-hit kinetics with x rays. Probes of the meristem that utilize different qualities of radiation offer a set of experimental tools for altering the genetic constitution in meristems in order to study further the genetic control of plant development.

Steffensen[21] has presented an elegant analysis, stemming from earlier work of Stein and Steffensen[22,23] in which chlorophyll deficient cell lineages, arising from dominant marker deletions induced by radiation, were used to reconstruct cellular development from the shoot apex of maize. Various lines of evidence led to the conclusions[21] that: "Cells in the lower portion of the apex give rise to daughter tissue much earlier in development than cells in the crest of the apex. The closer a cell is to the dome, the later daughter tissue will appear in development. Apparently the cells at the crest of the apex contribute the daughter tissue to nearly half of the upper part of the maize plant. Although there is some variation and chance in apical development, the major features of apical growth are well ordered and relatively predictable."

Two general tenants of the genetic control of development are that, with few exceptions, all cells of an individual contain the complete genome, and that differentiation involves differential gene activity. It is becoming increasingly apparent that large differences in size of nucleus and chromosomes as well as differences in nuclear protein content exist in different tissues and organs, or in cells of the same tissues during different stages of development. Experimental evidence in plants for such variations, as well as references to pertinent literature, were presented in recent papers by Bennett and Rees[3] and Bennett.[2] In each of three species studied, namely Secale cereale (rye), Allium cepa (onion) and Vicia faba (broadbean), they found that with the onset of growth after germination the size of metaphase chromosomes in meristematic cells first increases to a peak then decreases. The variation is associated with changes in nuclear protein and RNA content, while the DNA remains constant in quantity. Bennett stated that the increase in chromosome volume and mass could be brought about by increased gene transcription giving more mRNA, increased chromosomal metabolism not involving change in template activity, or differential removal of the products of either. Evidence is presented for the conclusion that changes in chromosome size reflect changes in metabolic activity of the meristem in which they occur and probably include changes in template activity.

In the experiments on maize reported here we know only that the size of the target, i.e., the distance from $\underline{Yg}_2$ locus to centromere, differs in interphase nuclei of three embryonic leaves. This would seem to imply that all regions of all chromosomes are involved in size variation. We have no measurements on DNA or protein; however, recent radiation experiments with Arabidopsis[6] seeds, using a similar genetic system for analysis, indicated that the chromosome as a whole is not the target, but rather, that the ultimate sites of radiation damage are of the order of size of single or double-stranded DNA.

The various observations discussed above[2,3,6] appear to be compatible with the interpretation that the nuclear proteins vary among cells according to differences in metabolic activity so that the constant amount of DNA undergoes different degrees of compacture, "packaging", or spacing in the chromosome of different organs or tissues.

The direct correlation in maize of leaf age with size of the nucleus and chromosomes suggests a further correlation of genetic activity with stage of leaf development. An analogy with the work of Bennett and Rees[3] may imply that maize embryonic leaf 3 was undergoing the highest metabolic and gene activity, and leaf 5 the least, at the developmental stage when the embryo became dormant in the kernel. Additional experiments are needed to determine whether the size of the nucleus is correlated more with developmental stage or more with a particular leaf throughout development. Do the nuclei of primordial leaf 5 become the same size as those of leaf 4 when they reach a comparable stage of development, or are the nuclei of leaf 4 always larger on an average than those of leaf 5? A preliminary examination of older leaves indicates that nuclei in different tissues of the same leaf may vary in size, and thus, different parts of a mature leaf would be expected to vary in radiosensitivity.

A problem for future studies is the possibility of relating nuclear size, DNA compactness, gene activity and organ or tissue specificity for some particular gene products. In a general discussion of gene activity and animal development, Schultz[10] concluded, "The primary molecular phenomenon concerned with these genetic activities is most probably the compaction and extension of the [DNA] template." Among recent observations on plant materials, a phenomenon which may bear on these problems is the occurrence, in suspensor cells of Phaseolus, of very large polytene chromosomes and nucleolar apparatus which are dispersed at optimal temperatures and compact at low or high ones.[7]

SUMMARY

Three sets of experiments are described. In the first, a 25 μ deuteron microbeam was used to mark regions in leaf primordia of dry maize seeds and in this way relative growth was measured for different leaf areas from seed meristems to mature plant organs. In the second, seeds irradiated with a wide range of doses of fission neutrons and of x rays were measured for various responses ranging from sublethal genetic effects to seedling lethality. Plant development, following "reprogramming" of the genotype by the treatment, proceeded along a highly predictable course to maturity, depending on dose and quality of radiation. In the third set of experiments, it was shown that the nuclei in different leaf primordia differ in size. On the basis of computed radiation target dimensions, this difference was also shown to involve distances along chromosome segments, suggesting differences in DNA compacture, and possibly in activity, in the different leaf primordia.

REFERENCES

1. Baker, C. P., Curtis, H. J., Zeman, W. and Woodley, R. G. (1961) The design and calibration of a deuteron microbeam for biological studies. Radiation Res. 15: 489-495.

2. Bennett, M. D. (1970) Natural variation in nuclear characters of meristems in *Vicia faba*. Chromosoma 29: 317-335.

3. Bennett, M. D. and Rees, H. (1969) Induced and developmental variation in chromosomes of meristematic cells. Chromosoma 27: 226-244.

4. Biavati, M. H., Rosenzweig, W., Rossi, H. H. and Miyanaga, I. (1963) The dependence of RBE on the energy of fast neutrons. III. Evaluations of radiation quality. Radiation Res. 19: 512-525.

5. Delihas, N. and Curtis, H. J. (1958) The relative biological effectiveness of fission neutrons for the production of acute mortality in mice. Radiation Res. 8: 166-180.

6. Hirono, Y., Smith, H. H., Lyman, J., Thompson, K. and Baum, J. (1970) Relative biological effectiveness of heavy ions in producing mutations, tumors, and growth inhibition in the crucifer plant, *Arabidopsis*. Radiation Res. 44: 204-223.

7. Nagl, W. (1970) Temperature-dependent functional structures in the polytene chromosomes of *Phaseolus*, with special reference to the nucleolus organizers. J. Cell Sci. 6: 87-107.

8. Rhoades, M. M. (1950) Meiosis in maize. J. Heredity 41: 59-67.

9. Rossi, H. H. (1966) Microdosimetry. IAEA Tech. Report Ser. 58: 81-95.

10. Schultz, J. (1965) Genes, differentiation and animal development. Brookhaven Symp. Biol. 8: 116-147.

11. Schwartz, D. and Bay, C. E. (1956) Further studies on the reversal in seedling height dose curve at very high levels of ionizing radiations. Amer. Naturalist 90: 323-327.

12. Smith, H. H. (1967) Relative biological effectiveness of different types of ionizing radiations: cytogenetic effects in maize. Radiation Res., Suppl. 7: 190-195.

13. Smith, H. H. (1971) Comparative genetic effects of different physical mutagens in higher plants. IAEA/FAO, Induced Mutations in Plant Improvement (Buenos Aires, 1970) (In press).

14. Smith, H. H., Bateman, J. L., Quastler, H. and Rossi, H. H. (1964) RBE of monoenergetic fast neutrons: cytogenetic effects in maize. In: Biological Effects of Neutron and Proton Irradiations, Vol. II, pp. 233-248, IAEA, Vienna.

15. Smith, H. H., Combatti, N. C. and Rossi, H. H. (1968) Response of seeds to irradiation with x rays and neutrons over a wide range of doses. IAEA Tech. Report Ser. 92: 3-8.

16. Smith, H. H., Curtis, H. J., Woodley, R. G. and Stein, O. L. (1962) The deuteron microbeam as a tool in botanical research. Radiation Botany 1: 255-268.

17. Smith, H. H. and Rossi, H. H. (1966) Energy requirements and relative biological effectiveness for producing a cytogenetic phenomenon in maize by irradiating seeds with x rays and monoenergetic neutrons. Radiation Res. 28: 302-321.

18. Smith, H. H. and Rossi, H. H. (1970) Relation between mutation yield and cell lethality over a wide range of x ray and fission neutron doses in maize. IV Intern. Congr. Radiation Res., Evian, p. 202 (abstract).

19. Smith, H. H., Rossi, H. H. and Kellerer, A. (In preparation).

20. Sparrow, A. H. (1965) Relationship between chromosome volume and radiation sensitivity in plant cells. In: Cellular Radiation Biology, pp. 199-222, Williams and Wilkins, Baltimore.

21. Steffensen, D. M. (1968) A reconstruction of cell development in the shoot apex of maize. Amer. J. Botany 55: 354-369.

22. Stein, O. L. and Steffensen, D. M. (1959) The activity of x-rayed apical meristems: a genetic and morphogenetic analysis in Zea mays. Z. Vererbungslehre 90: 483-502.

23. Stein, O. L. and Steffensen, D. M. (1959) Radiation-induced markers in the study of leaf growth in Zea. Amer. J. Botany 46: 485-489.

-DISCUSSION-

SAVAGE - Do you have any data on the cycle position for cells of leaves 4 and 5? If one were mainly in G_1 and the other predominantly in G_2, the loss due to aberrations would be quite different, involving essentially chromosome breaks in the former and chromatid in the latter.

SMITH - The chromosome segment volume differences do not suggest a difference in cycle stage between these two leaves.

CONGER, A. - There are extreme RBE differences present in your analyses. For example, the reversal of plant height has an RBE widely divergent from that of the Yg locus; it is difficult to relate the former to a chromosomal phenomenon.

SMITH - There is no difficulty because our experiments "end" where Dr. Haber's experiments begin. Seedlings measured for plant height reversal, are not gamma plantlets as they have mitotically active cells, so the effect could be due to genetic damage as it is with the Y_2 locus response.

HABER - Dr. Smith's RBE values for seed irradiation and our RBE values for seedling irradiation are in complete accord. Our interpretation is perhaps slightly different in that there is a high RBE for "cytological" components and a low RBE for "physiological" components.

THE CONFERENCE IN RETROSPECT

The purpose of this conference was to assess our current understanding of the structure and function of plant meristems and the program was arranged so as to give maximum opportunity for exchanges of information and discussion between participants. This summary attempts to bring out the salient points which resulted.

The morphology of the terminal root meristem is now well established, and several zones are readily distinguishable within the apical region by such characteristics as cell division rates and uptake of tritiated thymidine. A "quiescent center" is found in all roots so far investigated but its shape and size vary between species. Hitherto this region has been located by predominantly negative evidence and there is great need for some positive characterization. Such evidence would go a long way to solving the problem of its universality. Is it just the inevitable consequence of the spatial geometry of cells forming the meristem, or has it some functional significance such as the source of hormonal organizing gradients?

Radiation studies have shown that it can act as a reservoir of cells for meristem repopulation when need arises, and the mobilization of the quiescent cells appears to be extremely rapid under stress stimuli. This however is unlikely to be its primary function.

Increasing importance must be attached to the root cap, ultrastructural electron microscope studies of which have revealed hitherto unsuspected complexity. Cells of distinct types, and in some species of distinct origins, are found in this organ. Cells of the central core are concerned with the perception of gravity, and the transmission of such information back to the main meristem. The mechanisms of perception and transmission are still largely unknown. Cells of the outermost region are responsible for the production, accumulation and release of a mucopolysaccharide slime, which presumably acts as a lubricant, facilitating the passage of the root through soil.

Decapitation studies emphasize the complexity of the interrelationship of root and root cap, and should provide a profitable area for future research.

Morphological investigations of dormant roots have shown that heterogeneity of the cell population is already present in the embryo. Whilst the majority of cells appear to "rest" in G_1 this is by no means a universal rule and considerable variation occurs between species. This probably points to the existence of differing degrees of control over the maturation process. Comparative studies of root and shoot meristems suggest that differences in maturation may occur within the same embryo.

Even in those seeds with a pure G_1 population, heterogeneity still exists both with respect to the chromosome duplication status of individual nuclei, and to the radiosensitivity of the population.

These facts emphasize the need to know the cellular conditions obtaining within any meristem used for study, otherwise erroneous interpretation of experimental data is inevitable.

Studies in the very early initiation and development of root meristems may be made with lateral roots. Here again heterogeneity of cell cycle and division rates occurs, but in ways quite unlike those occurring in the developed terminal root. No quiescent center can be detected in very young lateral roots so this material offers the advantage of being able to study the genesis of this region, and indeed of the whole root apex organization. Such investigation is not possible in the terminal root since full organization is already established in the embryo.

The apparently random distribution of cell rates throughout the developing lateral root meristem makes it difficult to envisage some central control region within the meristem and it may well be that in very young roots individual cell control is the rule.

Culture methods for intact roots offer further advantages for studying the mechanisms which control cell division within the roots, both as to its initiation and cessation. However, there can be considerable problems in the interpretation of cell cycle kinetics particularly when the meristem is considered to be a single homogeneous unit -- which is frequently done. Any treatment involving perturbation of normal cell progression will give rise to erroneous quantitative estimates of such factors as mitotic index, labelling index, and yields of structural chromosomal aberrations.

Turning to the structure of the shoot apex we find that much less is known. This is partially attributable to the fact that these meristems are much more difficult to work with, and are not so amenable to the current techniques of meristem research.

The various morphological cell regions are not so obvious or well defined as in roots. In addition, the shoot apex is really a compound structure containing sub-meristems which give rise to leaves or floral parts, and the interaction between these sub-meristems will undoubtedly confuse the morphological picture.

The question of the "meristem d'attente" both as to its extent and position is far from being solved, as also is the question of its relation to the "quiescent center" of the root meristem. Methods are available for studying the surface growth of the living apex in ways which do not interfere with its subsequent development. These have established the fact that cell movement does occur over the surface of the meristem, but the relationship of these surface movements to the cell flow within the interior or corpus of the apex is not at present settled. This question is worthy of future pursuit.

There is also need for detailed comparative studies of shoot and root apex within the same species, where one has the same genetic constitution.

Another field which has been the subject of much investigation is that of the mechanisms which exercise overall control of development of meristems and plant organs. It has been well established that interaction between nutrition and hormonal action play a large part in this phenomenon.

Control occurs at at least two levels:

A) That which initiates, and stops development.
B) That which directs and determines the ultimate morphogenetic destiny.

One experimental mechanism for the initiation of meristematic activity is wounding, and for some forms of tissue differentiation (e.g., vascular repair) cell division appears to be prerequisite. Thus, the two control levels A and B may not be unrelated in some situations. The whole subject of the relationship of division to differentiation needs detailed consideration bearing in mind the observations that have been made on gamma-plantlets.

The possibilities of lateral transfer of hormonal substances across a node opens up new avenues for control, and raises many questions about the translocation, utilization and accumulation of hormones.

In the case of leaves and lateral buds, it may be that their ultimate fate is irreversibly determined much earlier than we have suspected, perhaps as early as the initiation in some species. It is difficult to see how this can arise as the result of generalized hormonal concentration gradients in the apical meristem. The findings are more consistent with localized genetic factors.

Ionizing radiation has proved to be a useful tool in probing the structure and control mechanisms of meristems, and of plant growth. Because of ease of handling, and the scoring methods available, seeds, particularly those of cereals, have been widely used.

From highly irradiated seeds, morphologically recognizable plants ("gamma-plantlets") may develop, which although incapable of cell division, nevertheless are physiologically competent for a wide range of cell functions. This indicates that a significant amount of pre-programming exists at the time of seed maturation.

Radiobiological experiments at lower, non-sterilizing doses have, in general, produced results consistent with our knowledge of the structural characteristics of meristems. In addition, some indication has been given of the sequential development of different meristems, and of the potential growth patterns of organs contained within the ungerminated embryo.

In spite, however, of all the efforts put into the study of meristems over the past decade, it is clear from the discussions at this conference, that we have only just begun to understand what meristems are all about. It is also clear that they will provide much research material for many years to come.

Morton W. Miller
Charles Carroll Kuehnert
John R. K. Savage

*This paper is based on work performed under contract with the U.S. Atomic Energy Commission at The University of Rochester Atomic Energy Project and has been assigned Report No. UR-49-1501.

AUTHOR INDEX
(Underlined numbers indicate Discussion)

SUBJECT INDEX

MIX
Papier aus verantwortungsvollen Quellen
Paper from responsible sources
FSC® C105338

If you have any concerns about our products,
you can contact us on
ProductSafety@springernature.com

In case Publisher is established outside the EU,
the EU authorized representative is:
Springer Nature Customer Service Center GmbH
Europaplatz 3, 69115 Heidelberg, Germany

Printed by Libri Plureos GmbH
in Hamburg, Germany